MONOGRAPHS ON
PHYSICAL BIOCHEMISTRY

GENERAL EDITORS

W. F. HARRINGTON, A. R. PEACOCKE

DIELECTRIC BEHAVIOUR OF BIOLOGICAL MOLECULES IN SOLUTION

BY

E. H. GRANT, R. J. SHEPPARD AND G. P. SOUTH

CLARENDON PRESS · OXFORD

Oxford University Press, Walton Street, Oxford OX2 6DP

OXFORD LONDON GLASGOW
NEW YORK TORONTO MELBOURNE WELLINGTON
IBADAN NAIROBI DAR ES SALAAM LUSAKA CAPE TOWN
KUALA LUMPUR SINGAPORE JAKARTA HONG KONG TOKYO
DELHI BOMBAY CALCUTTA MADRAS KARACHI

ISBN 0 19 854621 1

© Oxford University Press 1978

All rights reserved. No part of this publication may be reproduced, stored in a retrieval system, or transmitted, in any form or by any means, electronic, mechanical, photocopying, recording, or otherwise, without the prior permission of Oxford University Press

QH
506
.G65

Printed in Great Britain by
Thomson Litho Ltd, East Kilbride, Scotland.

PREFACE

The use of dielectric methods to study the composition of materials and the structure of their constituent molecules has a long history and numerous experiments had been carried out even before the beginning of this century. After the successful application of the technique to determine the dipole moment and other molecular parameters for simple solids and liquids attention was turned to biological materials in the 1930s. Despite the excellence of much of this work, progress was retarded by two factors, one experimental and one theoretical, both deriving from the high water content of substances of biological interest. Water and aqueous solutions strongly absorb radiowaves and microwaves and the weak field strengths encountered placed heavy demands on the older apparatus. On the theoretical side, the difficulties in interpretation of the dielectric data stem from the complexity of the system being studied. Water, the most common biological material of all, although a pure liquid of molecular weight low in comparison with that of other common biological substances, is nevertheless a liquid of great complexity. Moreover, any specific biological material, such as blood or muscle tissue, is a mixture of several different types of polar molecule, of which water is only one.

Because of recent progress in the development of stable sources and sensitive methods of detection, together with general improvement in the design and construction of specimen holders, it is now possible to obtain accurate values of the permittivity (dielectric constant) and conductivity of biological materials in most of the frequency regions of interest. The use of computers for the analysis of experimental data has brought a new rigour to the interpretation of dielectric measurements and has removed some of the ambiguities present in some of the earlier work. The computer has likewise played a significant part in further improving the theory of liquids and aqueous solutions by allowing structures to be calculated and models to be tested in a manner which would not hitherto have been possible. These advances have had the effect of carrying

PREFACE

the application of dielectric methods in biophysics one stage further, as well as opening up new horizons, and it has seemed to us that a book of this kind is therefore timely. As the title implies, we are concerned here with biological solutions rather than with tissues in general. This is because we judge that any trained scientist can now expect to be able to make dielectric measurements on liquids of biological interest and interpret the data with sufficient rigour to obtain useful unambiguous information at a molecular level. The same claim cannot yet be made for determinations carried out on solid or gel-like biological materials, although there has been progress in this area recently. The dielectric measurements that have so far been made on solid tissues can give information on tissue composition and, in particular, water content and the state of binding of the water molecules. Information of this type will be of increasing importance in such diverse areas as the treatment of cancer by hyperthermic methods and the evaluation of microwave radiation hazards.

I am very pleased that Dr. R.J. Sheppard and Dr. G.P. South were able to agree to my invitation to collaborate in the writing of this book. Their respective experimental and theoretical expertise, together with their considerable personal experience of the field, have resulted in the production of a text which is far greater in depth than anything I could have produced on my own.

Much of the work described in this book has involved myself or my co-authors and it is a pleasure to acknowledge the assistance of colleagues and collaborators with whom we have been associated. In particular I would like to thank Dr. T.J. Buchanan and Dr. H.F. Cook for introducing me to the field and for all their help in earlier days. I am greatly indebted to Professor H.P. Schwan for his continuing encouragement and support, and for the various occasions when it has been possible to work in his department at the University of Pennsylvania. Acknowledgement is also due to Dr. S. Takashima of the same department for numerous discussions. I would like to thank Professor R.E. Burge for providing the experimental facilities at Queen Elizabeth College and for his support of the research. A fairly recent area of concern is the dangers to personnel of

PREFACE

exposure to microwaves and I am indebted to Professor J.C. Gallagher of the University of Bradford for many discussions on the place of dielectric measurements in this important field.

During the writing of this book Dr. A.R. Peacocke, Editor of this series of Monographs, has made many helpful suggestions and we are grateful to him for his assistance. Finally, it is a great pleasure to thank Mrs. K.M. Grant for typing the manuscript and for general secretarial assistance.

September 1977   E.H.G.
Physics Department
Queen Elizabeth College
London

CONTENTS

1. **THE ROLE OF DIELECTRIC STUDIES IN BIOPHYSICS**
   1.1. General introduction — 1
   1.2. Dielectric studies in biophysics: historical survey and recent developments — 6
   1.3. Structure of biological molecules in aqueous solution: relationship with dielectric behaviour — 9
   1.4. Role of dielectric measurements in the study of water in biological systems — 12
   1.5. Biological and medical applications of dielectric studies — 17
   1.6. Dielectric studies as a tool in biophysics: résumé and outlook — 22

2. **DIELECTRIC THEORY**
   2.1. The nature of dielectrics — 26
     2.1.1. *Electric forces and fields* — 26
     2.1.2. *Polarization* — 29
     2.1.3. *Polar and non-polar dielectrics* — 32
     2.1.4. *Dielectric relaxation* — 35
   2.2. The static relative permittivity — 36
     2.2.1. *The meaning of static permittivity* — 36
     2.2.2. *The electric displacement vector* — 38
     2.2.3. *Theory of the static permittivity of fluids* — 39
   2.3. Time dependence and complex permittivity — 43
     2.3.1. *Relaxation and resonance* — 43
     2.3.2. *Dielectric relaxation and complex permittivity* — 44
     2.3.3. *Dispersion equations* — 47
     2.3.4. *Theory of frequency-dependent permittivity* — 50

3. **THE MEASUREMENT OF PERMITTIVITY**
   3.1. Introduction — 57
   3.2. Dielectric bridges — 58
     3.2.1. *Introduction and general theory* — 58
     3.2.2. *Low-frequency bridge measurements: the problem of electrode polarization* — 62
     3.2.3. *Medium-frequency bridge measurements* — 66
     3.2.4. *High-frequency bridge measurements: the problem of self inductance* — 70
   3.3. Coaxial line — 75
     3.3.1. *Introduction and basic theory* — 75
     3.3.2. *Higher modes* — 80
     3.3.3. *Design of a coaxial line cell* — 82
     3.3.4. *Methods of measurement not requiring a computer* — 84
     3.3.5. *Non-automated computerized techniques* — 86
     3.3.6. *Automated computerized techniques* — 87
     3.3.7. *Corrections for wall loss and ionic conductivity* — 91

3.4. Waveguides	93
3.4.1. *Introduction*	93
3.4.2. *Traditional measuring systems*	95
3.4.3. *Recent waveguide developments*	96
3.5. Other frequency domain techniques	99
3.6. Time domain spectroscopy	101
3.6.1. *Introduction*	101
3.6.2. *Basic theory*	102
3.6.3. *Single response methods*	107
3.6.4. *Multiple response methods*	113
3.6.5. *Conductive solutions*	118
3.6.6. *Conclusions on time domain measurements*	119

4. THE ANALYSIS OF EXPERIMENTAL RESULTS

4.1. General introduction	121
4.2. Graphical methods for preliminary analysis of data	121
4.3. Least-squares minimization	124
4.3.1. *Introduction*	124
4.3.2. *The Marquardt method*	126
4.3.3. *Practical details for a curve-fitting program*	129
4.4. The statistical analysis of data	130
4.4.1. *Introduction*	130
4.4.2. *Confidence intervals and the t-distribution*	130
4.4.3. *Correlation coefficients*	133
4.4.4. *Confidence contours*	134
4.4.5. *Choosing the best model — the F-distribution*	136
4.5. Practical considerations	137
4.6. Some functions encountered in the analysis of dielectric data	140

5. WATER AND SMALL BIOLOGICAL MOLECULES

5.1. Dielectric dispersion curve of water	145
5.1.1. *General features*	145
5.1.2. *The meaning of ε_∞*	148
5.2. Structural interpretation of the static permittivity of water	149
5.2.1. *General considerations*	149
5.2.2. *Interpretation of the static permittivity of water in terms of the dipole moment of the molecule in the liquid*	151
5.2.3. *Other ways of interpreting the static permittivity of water*	154
5.3. Interpretation of dielectric relaxation in water in terms of molecular motion	156
5.4. Dielectric properties of bound water	
5.4.1. *The δ-dispersion and its interpretation*	162
5.4.2. *The dielectric decrement near 1 GHz*	165
5.4.3. *Measurement of bound water from the radio-frequency dispersion*	167
5.4.4. *Calculation of hydration in solution from measured dielectric parameters: comparison with wet powders*	167

5.5. Dielectric behaviour of small biological molecules in solution: Macroscopic properties — 171
    5.5.1. *The nature of the problem* — 172
    5.5.2. *Determination of static permittivity and its use on a simple basis* — 174
    5.5.3. *Measurement of the dispersion parameters* — 176

5.6. Dielectric behaviour of small biological molecules in solution: Molecular interpretations — 181
    5.6.1. *Static permittivity* — 181
    5.6.2. *Dielectric relaxation in solutions of small biological molecules* — 184

6. PROTEINS AND LARGER STRUCTURES

6.1. Introduction — 189

6.2. The β-dispersion in protein solutions — 191

6.3. The deduction of molecular parameters for proteins — 195
    6.3.1. *Size, shape, and hydration* — 195
    6.3.2. *Dipole moments* — 207

6.4. Larger structures — 213
    6.4.1. *DNA* — 213
    6.4.2. *Lipoproteins* — 215
    6.4.3. *Viruses* — 216
    6.4.4. *Cells and membranes* — 217

BIBLIOGRAPHY — 221

AUTHOR INDEX — 229

SUBJECT INDEX — 234

1
THE ROLE OF DIELECTRIC STUDIES IN BIOPHYSICS

1.1. General Introduction

It is well known that a slab of material placed between the plates of a capacitor increases its capacitance by a factor called the dielectric constant or relative permittivity. Most substances used for this purpose have a dielectric constant lying between 2 and 10 and are good insulators. Such materials are termed non-polar because the constituent molecules do not bear a permanent dipole moment capable of rotating in an electric field. A non-polar molecule easy to visualize is that of benzene, which is in the shape of a regular hexagon with a CH group at each corner. This implies that the distribution both of positive and of negative charge is symmetrical with respect to the centre of the molecule and therefore that there is no molecular dipole moment. If one of the hydrogen atoms is replaced by chlorine to form monochlorobenzene the distribution of charge becomes asymmetrical and the molecule now bears a permanent dipole moment, i.e. it becomes polar. Owing to the interaction of the dipole moment with the electric field a polar substance has a dielectric constant which is larger than that of non-polar materials. The reason for this is explained in § 2.2.3. Another important difference between the two classes of substance is that the dielectric constant of a polar material is strongly dependent upon various physical parameters such as temperature, pressure, and frequency of applied field, whereas the electrical properties of a non-polar material are largely independent of them. Of particular importance is the variation of relative permittivity (ϵ') with frequency. For a polar substance ϵ' decreases with increasing frequency as the motion of the molecular dipoles becomes progressively unable to keep up with the changes in direction of the electric field. Accompanying this fall in permittivity is an absorption of energy by the medium from the field. For a polar material the term 'dielectric constant' is a misnomer and it is now considered preferable by most authorities to replace it by 'relative permittivity' or, as we shall do in this book, by 'permittivity', omitting any additional qualifications.

All biological molecules are polar but the value of the dipole moment (μ) depends very much on the size of the molecule and on the distribution of charge throughout it. A convenient way of expressing a molecular dipole moment is in terms of the Debye unit (D) which is named in honour of Peter Debye, who made a considerable number of contributions to the study of dielectrics in the first half of this century. A system of two charges each numerically equal to that of the electron, but of opposite sign, separated by a distance of 1 nm has a dipole moment of about half a Debye unit. On this basis the water molecule has a dipole moment of about 2D; for a large protein molecule such as haemoglobin (molecular weight 68 000) the magnitude of the molecular dipole moment is several hundred Debyes. Although we have here used SI units we have retained the use of the Debye unit for expressing the dipole moment of a molecule: the conversion factor between Debye units and SI units is given in § 2.1.3.

For a pure polar substance with only one type of molecule present, the value of the permittivity falls from one plateau to another as the frequency of the applied field is increased. With a solution where both solute and solvent are polar two dispersions would be expected. Simple solutions of biological interest fit in this category. The dispersion curve for a 20 per cent (200 kg m^{-3}) solution of haemoglobin in water is illustrated in Fig. 1.1. The permittivity falls from about 180 to about 65 as the frequency increases from 50 kHz to 50 MHz followed by a gradual decline until a frequency of near 500 MHz is reached. After this the permittivity again rapidly falls with increasing frequency and levels out at a value of between 4 and 5 at frequencies around 500 GHz. If we disregard the slight drop in permittivity at frequencies of a few hundred MHz, the figure shows the presence of two large dispersion regions. The one at low frequencies is due to the relaxation of the relatively large protein molecules; that at high frequencies is due to the smaller molecules. The lower curve in the figure is the dielectric loss (ε''), which is proportional to the energy absorbed from the field. The peak in ε'' occurs at the relaxation frequency (f_R), which is about 1 MHz for haemoglobin and 20 GHz for water. The reciprocal of $2\pi f_R$ is the relaxation time (τ).

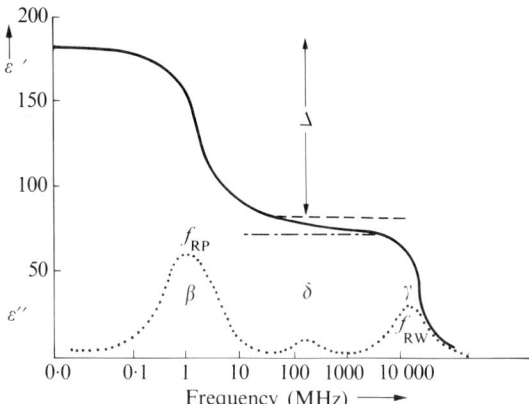

FIG. 1.1. Dielectric dispersion curve of an aqueous solution of haemoglobin. The curve shows three separate dispersion regions. f_{RP}: relaxation frequency of protein molecule. f_{RW}: relaxation frequency of water molecule. Δ: dielectric increment. The relaxation time, $\tau = (2\pi \times \text{relaxation frequency})^{-1}$. ----, β-dispersion extrapolated to high frequencies, -·-·- γ-dispersion extrapolated to low frequencies

The fall in permittivity throughout the dispersion (Δ) is proportional to the square of the molecular dipole moment; the value of τ is related to the size of the molecule in solution; and the magnitude of the temperature variation of τ is determined by the nature of the intermolecular bonding. Thus it may be seen, in very general terms, how information at a molecular level can be gained from a study of dielectric behaviour.

Not all biological material is as simple in composition as an aqueous protein solution. A piece of tissue contains much membranous material owing to the presence of cells and also many ions, as well as water and the various macromolecules, which may themselves range from lipoproteins and nucleic acids of molecular weight around one million down to amino acids of molecular weight about one hundred. Fig. 1.2 shows the dielectric dispersion curve of muscle tissue, which, as has been pointed out by Schwan (1957, 1974) can be broken down into three principal regions which he termed the α-, β-, and γ-dispersion.

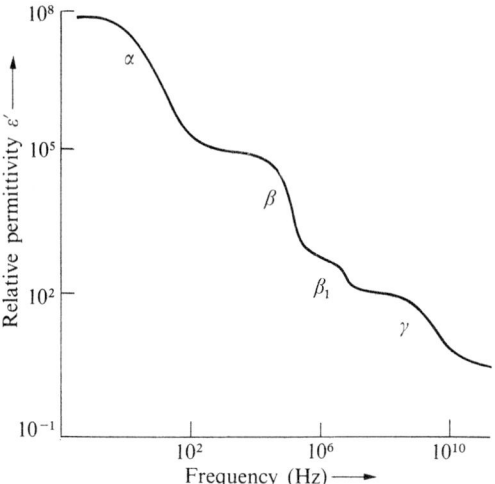

FIG. 1.2. Relative permittivity of muscle tissue (from Schwan 1974)

The α-dispersion is the least well understood of the three but it could be due to the relaxation of counterions surrounding the charged cell membranes. Another possibility to account for its origin is the migration of ions through the holes in the membrane. The shape and size of the α-dispersion is very dependent upon the metabolic state of the tissue and the study of it frequently lies more in the realm of electrophysiology than of biophysics. In this book we shall refer to the α-dispersion only when reproducible and biologically stable systems are being considered. An example of this is the lipoprotein study (§ 6.4.2) where an α-dispersion is described which has a counterion relaxation as its origin.

The β-dispersion can be subdivided into two parts (Fig. 1.2). For reasons of clarity these two parts are shown separately, but in practice they may be less obviously distinct. The major part of the fall in permittivity (from several thousand to between one hundred and two hundred) is due to the inhomogeneity of the material and is an example of the Maxwell–Wagner effect. The mathematics of this mechanism of polarization is given in § 6.4.4 but a qualitative description is as follows. When a heterogeneous medium is placed in an electric field an accumulation of charge will rapidly build up at the various

boundaries separating regions of different permittivity and
conductivity. A finite time is taken for the charges to reach
equilibrium and the effective permittivity and conductivity
of the medium as a whole will depend upon how this time compares with the periodic time of the field. These considerations
predict a variation of permittivity with frequency of the required form. Study of the β-dispersion can give information
concerning the structure and width of the cell membranes
(§ 6.4.4). The fall in permittivity from 100–200 to below 80
may be referred to as the β_1-dispersion and is due to the rotation of biological molecules other than water. The protein
dispersion illustrated in Fig. 1.1 is an example of the β_1-dispersion. In this book we are concerned principally with the
behaviour of biological molecules, rather than that of larger
biological structures so we shall omit the suffix and refer to
the dispersion of the solute molecule simply as the β-dispersion.

The γ-dispersion is due to the relaxation of the water
molecules and is characterized by parameters which are close to
those appropriate to the dispersion of pure water. The chief
difference is in the value of the low-frequency permittivity,
which would be close to 80 for pure water at 20°C. For a biological solution the value of this parameter is less than 80
by an amount which is roughly proportional to the quantity of
solute material present. This lowering of the permittivity
below the value of pure water can be used to obtain information concerning the water immediately adjacent to the solute
molecules.

Although not shown explicitly in Fig. 1.2, there frequently
appears to be a small dispersion occurring between the β- and
γ-regions. This small dispersion region (the middle of the
three dispersions shown in Fig. 1.1) has now been observed in
the aqueous solutions of several different proteins and is
referred to as the δ-dispersion. Most workers in the field
consider the cause of the δ-dispersion to be the relaxation of
the water molecules in the immediate environment of the biological macromolecules. Although other mechanisms cannot be
entirely ruled out, our view is that it is highly probable that
this so-called bound water, or water of hydration, is respons-

ible for the δ-dispersion; we therefore consider that the bound water in a biological material may be studied from observations of the δ-dispersion. There are many other ways of studying hydration, one of which has been referred to in the previous paragraph. Bound water is of great biological importance and is discussed in more detail in § 1.4; the study of the quantity and nature of bound water by dielectric methods is covered in § 5.4.

1.2. Dielectric studies in biophysics: historical survey and recent developments

It is frequently some years before a technique which has proved successful for the study of non-biological materials can be used in equivalent studies on biological substances. There are several reasons for this delay. On the experimental side there is the problem of obtaining reproducible samples and the presence of water and ions in varying quantities. The complexity of the specimen may demand the making of a very large number of measurements to attain a limited objective and even then an unambiguous interpretation of the measurements may prove difficult. When a series of accurate results has been obtained and fitted to an appropriate equation, there still remains the task of interpreting the macroscopic parameters at a molecular level and, owing to their complexity, this is likely to be more difficult for biological molecules. However, once the necessary experimental and theoretical advances have been made the technique then becomes viable for use with biological materials and, in some cases, may be applied with outstanding success. A good example of this is the use of X-ray diffraction methods in the 1950s and 1960s to unravel the structures of DNA, haemoglobin, and myoglobin. The technique of X-ray diffraction had been introduced and used successfully for the study of elementary inorganic molecules like sodium chloride some forty years before equal success was achieved with complex biological molecules.

A similar situation occurred in the field of dielectric measurements. Although there are numerous reports of permittivity and dielectric loss measurements being made before the turn of the century, the materials studied were simple inorganic substances. It would be both difficult and unwise to try to identify the first piece of dielectric work carried out with biological materials, but the research achievements of Wyman and Oncley and their co-workers during the 1930s must rank amongst the first large-scale efforts in this field. Wyman used resonance methods at frequencies of a few hundred MHz to measure the static permittivity of the aqueous solutions of many small biological molecules including various amino acids, a few peptides, and urea. Recent work on the same materials confirms the results of Wyman and vindicates his experimental technique. Oncley measured the dielectric dispersion (the β-dispersion) in thirteen protein solutions and thereby deduced the dipole moment size and shape of the molecules in solution. The measurements were carried out using bridge methods operating in the frequency range of about 100 kHz–10 MHz. Bearing in mind the problems of reproducing the protein solutions which Oncley used, it is found that recent bridge measurements confirm Oncley's work. More specific reference is made to the work of Wyman and Oncley in Chapters 5 and 6 respectively. Other examples of early work are Bateman and Potapenko (1940) and Conner and Smyth (1942).

The development of microwave sources during the second world war extended the frequency range over which dielectric measurements could be made and this was put to good use in the late 1940s and early 1950s when the first measurements on the γ-dispersion appeared (Buchanan, Haggis, Hasted, and Robinson 1952; Cook 1952). At the same time Schwan and his colleagues were considering the problems of permittivity measurements on biological materials at frequencies below 100 kHz. The difficulties here lie in the measurement of a small dielectric term in the presence of a large conductive term and therefore a bridge capable of high resolution is required. Such an instrument was developed by Schwan and Sittel (1953).

Owing to the availability of bridges and to the introduction of transmission line methods specifically devised by Buchanan (Buchanan 1952; Buchanan and Grant 1955) for measuring the permittivity of conductive liquids, a fair number of publications appeared in the 1950s describing dielectric work on biological materials. The bridge measurements covered the range from a few kHz to near 100 MHz and the transmission lines operated from about 500 MHz—25 GHz. Subsequently the high-frequency limit of bridges was moved upwards and the low-frequency limit of coaxial line apparatus downwards; this led to the discovery of the δ-dispersion for certain protein solutions in the late 1950s and the early 1960s.

During the same period progress on the theoretical side had been much slower and the extensions and improvements to experimental techniques had not been matched by similar developments in dielectric theory. However, as we show in Chapter 6, the model used by Oncley to interpret his results on protein solutions requires little modification, at least in the case of globular proteins. The dipole moments calculated by Wyman for the smaller biological molecules were shown at the time to be consistent with the simple picture of the unimolecular rotation of individual dipoles of moment equal in value to the electronic charge multiplied by the length of the molecule. Recent work confirms this picture but requires the introduction of a factor to account for intermolecular attractions. Also the values of dipole moment calculated by Wyman are shown to be only a few per cent too low (§§ 5.5.2 and 5.6.1).

As with other branches of research, the introduction of computers (in the early 1960s) made a very big impact on dielectric studies. On the theoretical side it became possible to estimate the dipole moment of a protein molecule from the known co-ordinates of its constituent atoms. This method, in the case of myoglobin, is explained in § 6.3.2 where it is shown that the agreement between this calculated dipole moment and the experimental value is excellent considering the complexity of the system and the assumptions made. Not only is the absolute value of dipole moment correctly predicted, but the observed experimental differences of 20D between the dipole moments of horse and whale myoglobin, owing to the fact that 19 out of 153 amino acid residues (§ 1.3) are different, also follows directly from the calculations. This success is worth emphasizing because it shows that the dipole moment obtained from the dielectric dispersion curve is a parameter of molecular significance and not merely an empirical quantity.

The introduction of computers has enabled many experimental techniques to be automated and this has reduced the effort required for experimental investigation and has enabled the accuracy of the results to be increased (§ 3.3.6). Furthermore modern computer methods of data analysis enable the best possible information to be obtained from the experimental data (§§ 4.3 and 4.4).

With regard to experimental techniques for measuring the permittivity of biological materials, various new ideas (other than those related to the introduction of the computer) have been introduced in recent years. One that could have far-reaching consequences is measurement in the time domain rather than in the frequency domain, such techniques being known as 'Time Domain Spectroscopy' (TDS) and 'Time Domain Reflectometry' (TDR).

1.3. Structure of biological molecules in aqueous solution: relationship with dielectric behaviour

The most abundant biological substance is water and this unique material is discussed separately in the next section. As we shall see there and in Chapter 5, the relationship between the dielectric behaviour and the structure of water is complicated and not yet fully understood. For some of the larger biological molecules the interpretation of dielectric data in terms of structure is easier, paradoxical though this may appear at first sight. To help understand some of the principles involved it is worthwhile to consider the family of molecules represented by amino acids, peptides, and proteins. To obtain a comprehensive description of the structure and function of these molecules the reader should consult a textbook on biochemistry or molecular biology, such as Haggis, Mitchie, Muir, Roberts, Walker, Blow, and Sheinin (1974).

An amino acid has the characteristic form $H_2N.CHR.COOH$ where R represents a chemical grouping. In solution the molecule ionizes so that the amino group becomes H_3^+N and the carboxyl group COO^-, thus forming a zwitterion of dipole moment equal to the product of the electronic charge and the distance between the two groups. There are about 20 naturally occurring amino acids, the simplest of which is glycine where R comprises one hydrogen atom. Although the chemical grouping represented by R increases in complexity for the bigger amino acids the dipole moment should remain roughly the same, according to the zwitterion concept, provided that the amino group and carboxyl group are attached to the same carbon atom (the α-carbon atom). Moreover, the value of this dipole moment should be in good agreement with that calculated from known bond angles and interatomic distances. This expectation is borne out by experiment and thus vindicates the simple assumptions made. If R is a methyl group then two possibilities arise: the amino group can be attached either to the carbon atom joined to the carboxyl group (as above) or the amino group and carboxyl groups can be attached to different carbon atoms, in which case the separation of the groups, and so the molecular dipole moment, would be greater. Hence the two molecules, although chemically similar, would be expected to behave differently in a dielectric sense. The two molecules of this example are α-alanine and β-alanine, and it is known from dielectric measurements that the dipole moment of the β form is greater than that of the α form by the expected amount. If the simple zwitterion picture is correct then the dipole moment of an amino acid should progressively increase with the length of the molecule: an example is ε-aminocaproic acid (an important blood clotting agent) which has five CH_2 groups between the amino and the carboxyl groups and a correspondingly big dipole moment nearly twice that of an α-amino acid. The measured relaxation times of α- and β-alanine are the same (unlike their dipole moments), so it is the physical size of the molecule, rather than the charge distribution, which determines its rate of rotation. The dipole moments and relaxation times of amino acids are considered in § 5.6.

When two amino acids join a peptide is formed, with the elimination of one water molecule. Thus alanine and glycine form either alanylglycine $NH_3^+.CH.CH_3.CONH.CH.H.COO^-$ or glycylalanine, which is obtained by interchanging the methyl group with the hydrogen atom. The dipole moments and relaxation times of these two molecules are equal, as would be expected according to the elementary principles outlined in the previous paragraph.

When several amino acids join to form a more complicated peptide the situation becomes more involved owing to the presence of charged side groups and to the non-linearity of the chain. Thus a protein molecule such as haemoglobin has nearly 600 amino acid residues in four separate chains, each of which is composed partly of the well-known α-helix; moreover all four chains are folded in a complicated manner (Haggis *et al.* 1974). In such a case it is a major exercise to establish the relationship between the dipole moment and the structure, and we will postpone further discussion of this type of problem until Chapter 6. However, the relaxation time of around 100 ns does accord with what would be expected for a molecule of this size and, in fact, a roughly linear relationship holds between relaxation time and molecular weight for biological molecules ranging from water to protein molecules of molecular weight up to several hundred thousand. Most common biological molecules, and certainly those which have been intensively studied (by dielectric methods or otherwise), seem to fall into two categories. There are the smaller biological molecules, of molecular weight less than a few hundred, and macromolecules of molecular weight greater than about ten thousand. In the former category occur urea, the amino acids, and peptides, while the latter group contains proteins, nucleic acids, and lipoproteins. There are, however, a few important biological molecules of intermediate molecular weight, the investigation of which would repay the effort involved. One example is ATP (adenosine triphosphate) which is responsible for the storing of chemical energy within the cell. ATP has a molecular weight of 501. Measurements which we have carried out show that it has a dipole moment of around 30D and a relaxation frequency of 50 MHz (Roberts 1977, Roberts, Dawkins, Sheppard, and Grant 1978), which falls in one of the more difficult frequency ranges from the standpoint of obtaining high accuracy in the measured values of ε' and ε'' (Chapter 3). It is therefore rather fortunate that there are only relatively few interesting biological molecules of molecular weight between a few hundred and a few thousand.

For all the molecules considered in this section and, indeed, for all known biological molecules which exhibit dipolar rotation, the dielectric increment in aqueous solution (Δ) is proportional to the square of the molecular dipole moment. Biological molecules are well behaved in this sense, as is explained in Chapter 2, but it is worth emphasizing that this relationship should not be assumed too readily. It arises from the molecules rotating singly and in a similar environment which, in this case, is produced by the surrounding water molecules and by the formation of hydrogen bonds between them and the solute molecule in question. In contrast a lack of proportionality between Δ and μ^2 is observed in the dielectric behaviour of the liquid hydrides HCN, H_2O, HF, and H_2S, whose permittivities, as liquids at 20°C, are 115, 80, 65, and 4, but whose molecular dipole moments are 2.93, 1.84, 1.94, and 0.95D respectively. The precise reasons for this deviation from the general pattern are complex: an important contributory factor is that HCN rotates as a linear trimer while HF forms ring structures consisting of several molecules each. We presume that the molecules of H_2O rotate individually rather than in clusters (§ 5.3) but the important difference between water and H_2S, as far as dielectric behaviour is concerned, is that water forms hydrogen bonds but H_2S does not.

1.4. Role of dielectric measurements in the study of water in biological systems

Roughly 70 per cent of the human body is composed of water. This is an average figure, the amount of water present in each tissue varying from over 90 per cent in blood plasma to about 30 per cent in fatty tissue. Owing to the large abundance of water in the body and its unique properties, it might be expected that a considerable amount of research effort would have been spent on studying its structure and function in biological systems but, in fact, the reverse is the case; the water present in biological systems had largely been taken for granted until about fifteen years ago. Even now water is not given the attention which it deserves, as the study of most standard textbooks on medicine, biochemistry, or biology will show; it is only in the more specialized books such as the

THE ROLE OF DIELECTRIC STUDIES IN BIOPHYSICS

comprehensive treatise on water edited by Franks (1972) that the essential role of water in living systems is acknowledged. In recent years many research programmes have been initiated to study the behaviour of water in living tissue and much of this activity can be traced to a conference entitled 'Forms of water in biologic systems' organized by Saunders and Flynn and sponsored by the New York Academy of Sciences in 1964. At this meeting it was pointed out forcefully that water is essential for the initiation and preservation of normal metabolic activity and is not merely a useful liquid for filling up the gaps between the other (perhaps more obvious) tissues.

It is worth recalling some of the physical properties of water which make it such an important agent in living organisms. Its specific heat is higher than that of any liquid other than ammonia, its latent heat of vaporization is twice as high as that of any other known liquid, and its latent heat of fusion is also very large in relation to other liquids. These properties ensure good temperature stability in the animal or plant which may be exposed to sudden heat loss or gain owing to internal or external processes. The surface tension of water is the highest of any known liquid and this leads to the high capillary rise of water in plants and trees which is an essential factor in the transference of nutrients from the soil. Another important property of water is the well-known anomalous behaviour of its density, which exhibits a maximum at 4°C. In cold climates this causes freezing to occur on the surface of ponds and lakes and allows the presence beneath it of the liquid water necessary for the maintenance of marine life.

All this has been known for some time, but it is only recently that any understanding at a molecular level of why water is so important in biosystems has been achieved. It is now realized that when a biological macromolecule is present in an aqueous environment the water molecules play an important part in determining the structure and properties of the macromolecule and do not merely form an inert continuum in which it resides. There are several examples of this. The double helical structure proposed by Watson and Crick (1953) is well known, but what is less well recognized is the fact that this structure owes its stability, at least in part, to the presence of the neighbouring water molecules in the local environment, an observation which was made by Jacobson (1953) only a few months after the appearance of the original paper by Watson and Crick. Thus it is the *hydrated* DNA molecule which is the entity of biological importance, and the water molecules must be regarded as an integral part of this structural unit. Another biological molecule of great significance and importance is cholesterol; an excess of cholesterol in the body exhibits a strong correlation with coronary heart disease. It has a strong affinity for water and forms such a complete structural unit with it that it acts as a better substance for nucleating ice than silver iodide. Another interesting example is collagen, an important constituent of muscle, which is a protein molecule made up of three polyproline helixes. Proline is an amino acid molecule in the form of a closed pentagonal ring bearing CH_2 groups on three apices and with a CHCOO group and an NH_2 group on the other two apices. The helical chains are unable to form the intra-chain hydrogen bonds characteristic of the α-helix and it has been proposed by Berendsen and Migchelsen (1965), on the basis of NMR studies, that the stability of the collagen molecule is effected through the chains of water molecules running parallel to the direction of the helix. There are numerous other examples of important interactions between water and biological molecules but the factors determining the shape of a globular protein in aqueous solution are of especial interest. Such a molecule folds so that the hydrophobic groups occur on the inside of the molecule while the hydrophilic groups appear on the outside and make bonds with the numerous water molecules in the immediate vicinity.

The water molecules referred to in the above examples are in close proximity to macromolecules and will be influenced by the strong electric fields produced by them, so it would be expected that certain physical properties of water in *this* state may differ from those of ordinary water. Consequently, by selecting an appropriate experimental method, it should be possible to study the properties and behaviour of water in the immediate neighbourhood of biological macromolecules. As the dielectric technique is one of these methods, the nature of this 'structured' water warrants some further discussion at this point (see Chapter 5 for more specific details).

That 70 per cent of the human body is water, much of which is intimately involved with proteins, nucleic acids, and other large molecules, appears on first consideration to be in conflict with the rigid and typically solid characteristics of the body tissues. This proportion of solid material to liquid corresponds to 100 g of sugar in 200 ml of water, which sould simply form a solution having all the mechanical characteristics typical of a liquid. By comparison, if only 10 g of gelatin powder is mixed with the same volume of water a stiff jelly is obtained. Thus the effect of the gelatin has been to give the water a rigid structure having mechanical properties more like those of a solid than a liquid. Since this results from the binding of the water molecules to the gelatin, the water is referred to as 'bound' water or 'water of hydration' and, in this case, is recognized by the rigidity possessed by an otherwise fluid system. The presence of bound water in the body likewise enables it to exhibit more solid-like characteristics than would be expected for a biological system having an aqueous content of 70 per cent.

Owing to the nature of the strong forces between the bound water and its neighbouring macromolecules, it is expected that the relaxation time should be longer than that of a water molecule existing in pure liquid water, which may therefore, in contrast, be termed 'free' water. In other words the water–solute bonds will be broken much less often than the water–water bonds in pure water, and the dielectric dispersion curve of bound water should therefore occur at a lower frequency. This is the basic principle underlying the use of dielectric methods for investigating the quantity and nature of the bound water present in a biological material. The method suffers from the weakness that the investigator is looking at the properties of a small quantity of bound water against the background effects of all the biological macromolecules and free water present in the same system. However since, by definition, bound water cannot be isolated from the other components present in the system, this drawback applies to any other technique for investigating its behaviour whether it be dielectric dispersion, self-diffusion, sedimentation velocity, sedimentation equilibrium, osmotic pressure, calorimetric methods, low-angle X-ray scattering, or nuclear magnetic resonance (NMR).

The advantages and disadvantages of these various methods have been reviewed by Ling (1972), but we must point out that it is a mistake to expect more than rough agreement between the values of the bound water parameters calculated for a given biological specimen using these different techniques. This is because the definition of 'bound' water will vary with the technique. Thus referring to Fig. 1.1, the bound water calculated from the measured relaxation time of the β-dispersion is the water which has no translational freedom regardless of its rotational mobility (§ 5.4.3). The amplitude of the δ-dispersion measures the water present in the system which has a relaxation time about one hundred times greater than that of free water molecules. The depression of the permittivity at the high-frequency end of the δ-dispersion below that of pure water is determined by the amount of water which is unable to turn in the electric field at a frequency near 1 GHz (§ 5.4.2). The bound water measured in NMR studies is obtained from changes in the spin—spin relaxation time and from line-width broadening (for a survey of methods, see Kuntz and Kauzmann (1974)). There is no reason to expect the agreement between these estimates of bound water to be anything better than qualitative. So the absolute value of the hydration obtained by any particular method for a given biological solution is somewhat limited in its usefulness. On the other hand, the *comparative* estimates of the quantity of bound water present in different systems made by the same measuring technique may be of great significance. Examples of this are shown in §§ 1.5 and 6.4.2.

As rightly pointed out by Forster and Minton (1972) some of the claims made for the quality of the information obtained from studies of the δ-dispersion during the middle 1960s were over-optimistic. Recent improvements in the accuracy of measuring permittivity and in the preparation of samples, plus the advantages gained from the use of the computer, referred to earlier in this chapter, have considerably reduced the errors involved and make this method of measuring hydration competitive with all the others which are available. Very recently it has been shown that the δ-dispersion can be subdivided into two components (Essex, Symonds, Sheppard, Grant, Lamote, Soetewey, Rosseneu, and Peeters (1977)).

In this section we have concentrated attention on the study of the properties of bound water and their measurement; the behaviour of free water has been largely ignored. There are two reasons for this. One is that the biological importance of free water in bulk has been discussed adequately elsewhere and the second is that numerous other texts have appeared describing the dielectric behaviour of free water. This is summarized in §§ 5.1–5.3 and is used as a way of introducing the concepts involved in relating dielectric properties to structure. For a comprehensive review of protein hydration, readers are referred to Kuntz and Kauzmann (1974).

1.5. Biological and medical applications of dielectric studies

Dielectric studies have provided a great deal of information on the structure and properties of biological molecules and by thus contributing to the total body of knowledge, have eventual biological and medical application. However there are a few specialized areas where the applications are more direct and we shall refer to some of them in this section. Two categories will be considered: the first is the dielectric investigation of a biological material which has produced information of direct application to a clearly identifiable problem of medical and biological importance; the second concerns the situation when the known or suspected dielectric properties of a biological material are such as to produce interactions with radio waves or microwaves which can be of direct clinical or biological significance.

One of the strengths of the dielectric method in relation to the study of biological material is that the measurements can be made on aqueous solutions or suspensions. Hence systems can be studied in conditions which approximate to the *in vivo* state and the effect of the water molecules is included in the investigation.

One interesting problem concerns the mechanism of action of

certain anaesthetic agents. It is well known that chloroform, nitrous oxide, carbon dioxide, cyclopropane, argon, and xenon produce anaesthesia and the question to be answered is how such inert substances with such limited ability to take part in chemical reactions can initiate a chain of processes which gives rise to unconsciousness. One answer has been provided by Pauling (1961) and discussed by Catchpool (1965). According to this theory the one type of interaction in which these inert substances can take part is the formation of crystalline hydrates, i.e. areas of structured water are capable of being formed in the neighbourhood of the molecules of the anaesthetic agent. This formation of hydrate microcrystals in the encephalonic fluid increases the network impedance and thus produces narcosis. The conductivity of ice is known to be less than that of pure liquid water, so it is reasonable that the conductivity of the water of hydration should be less than the water present before the anaesthetic agent is introduced, i.e. the presence of the inert material increases the tissue impedance through decreasing the conductivity of the water content. According to this picture lowering the temperature of the brain should increase the probability of narcosis because of the greater likelihood of bound water formation. Lowering the pressure should have the same effect, again owing to the greater probability of the formation of hydrated water. Both of these effects are observed clinically.

One way of testing this theory would be to measure the effect on the dielectric properties of a biological solution after an inert gas has been dissolved in it. An experiment along these lines was carried out by Schoenborn, Featherstone, Vogelhut, and Susskind (1964), who measured the permittivity of 9 GHz of various solutions of haemoglobin containing dissolved xenon. Their results were consistent with an increased bound water content owing to the presence of the xenon and it would be worthwhile repeating the experiment over a wide frequency range in order to discover more information about the nature and the amount of the bound water present.

A different kind of problem which is currently receiving attention is the evaluation of the structure of lipoproteins present in blood serum. Several possible models have been suggested and two are of current concern. The first of these is the lipid core model in which the lipoprotein molecule is regarded as sphere of lipid surrounded by a shell of protein. The alternative proposal is the lipid bilayer model, which requires a central protein core and an external shell of protein with a layer of lipid between them. Dielectric measurements carried out by Essex provide evidence which firmly favours the lipid bilayer model and is inconsistent with the lipid core model. A brief reference to this work is given in § 6.4.2.

A useful application of dielectric studies in biology is in the measurement of membrane thickness. Most cell membranes have a capacitance of around 1 µF cm^{-2}, which corresponds to a membrane thickness of the order of 10 nm. This can be established from permittivity measurements made at radio frequencies and has been used to investigate lens material (Pauly and Schwan 1964) and suspensions of the bacterium *E. coli* (Carstensen 1967). Other examples of this kind of work are given in § 6.4.4. In the past few years dielectric measurements have produced information which appears to have the possibility of being applied directly to clinical problems. The use of permittivity studies to measure the difference between the quantity of serum lipoprotein bound water in healthy persons and those with genetically determined heart disease is referred to in §§ 5.4.2 and 6.4.2. Another study which could have important applications is due to Nordqvist, Arwin, and Lundström (1975). In this work the cerebrospinal fluid taken from patients with essential senile dementia was shown to have different dielectric properties from that present in healthy individuals. Both this and the lipoprotein study lead to the possibility of using the measurement of dielectric properties as a diagnostic tool. It is sometimes useful to be able to detect qualitative changes in biological material without the necessity of having to provide a rigorous interpretation of what has happened. An example of the use of dielectric measurements in this category is due to Jason and Lees (1971), who used the variation in permittivity with time of tissue taken from fish as an indication of its freshness, and a similar application is described by Kent (1977) who shows how the water content of frozen fish is related to its microwave attenuation. Another interesting use of dielectric measurements in biology is the use of permittivity determinations to study intracellular water in the metabolism of the embryo of the brine shrimp *Artemia salina* (Clegg 1978) and a further example of the application of permittivity measurements on biological materials is the use of microwaves to measure the water content in wool (Das 1972; Das and Smith 1972).

We shall now consider in more detail examples of the knowledge of the dielectric properties of tissue being used to evaluate the possibility of solving problems of biological or medical importance.

Radio waves and microwaves are used routinely for producing heat in tissues in the treatment of various diseases by the methods of physical medicine. The energy absorbed by a given tissue depends upon the permittivity and conductivity at the frequency of the radiation being used, and if these electrical parameters are known it is possible to build a 'phantom' body of the appropriate shape in order to measure the isopower lines and hence evaluate the distribution in temperature throughout the medium. A possible phantom material used is a mixture of water and dioxan containing dissolved potassium chloride. By varying the concentration of the three substances, the required combination of permittivity and conductivity can be produced. The permittivity and conductivity of various tissues have been measured by Schwan and Li (1953) and by Cook (1951). Phantom materials can be made using these data (Johnson and Guy 1972).

Recently consideration has been given to the idea of using radio waves and microwaves in the treatment of malignant disease. The employment of hyperthermia to cause tumours to regress is an old idea and in the past the procedure has been to raise the temperature of the whole of the body of the patient using some appropriate method such as a wax bath. However, as pointed out by Har-Kedar and Bleehen (1976) in a recent review, the use of an external beam is a more attractive idea because the heat can be localized in the area of interest. Moreover, if it could be shown that there were frequencies, or a range of frequencies, where malignant cells absorb heat differentially at the expense of normal cells, the technique of radio wave or microwave therapy would be particularly worthwhile. There are reasons for believing that the dielectric behaviour of normal and malignant cells may be sufficiently different to be optimistic about this approach. It has been shown by Bather, Webb, and Cunningham (1965) that dessication may be classed with other carcinogenic agents and that the water content of malignant cells is less than those which are normal. Nuclear magnetic resonance measurements of Gordon, Mallard, and Philip (1976) and Medina, Hazlewood, Cleveland, Chang, Spjut, and Moyers (1975) suggest that there are differences in the relative amounts of bound and free water in the two types of cell, and Burger and Noonan (1970), by studying the agglutinin sites on the surface of a tumour cell, show that the membrane surface is different from that of a normal cell. Thus there is evidence to suggest that water is bound differently in malignant and normal cells but it is too early to say what this difference is. Differences in the permittivity and conductivity of bound and free water cause different amounts of energy to be absorbed by the two states of water (de Loor 1973; Grant, Sheppard, and South 1975). We therefore conclude that although a considerable amount of work will have to be done before the precise differences between the water structure in normal and malignant cells are understood, the method of external beam hyperthermia offers encouraging possibilities in the treatment of cancer.

The use of microwaves or radio waves for the production of heat is a very well established phenomenon in industry but its employment in medicine and biology is somewhat more recent and, at present, more limited. An interesting biological application is the use of electromagnetic waves of frequencies in the range 10–100 MHz to kill insects which destroy wheat and other crops (Nelson and Charity 1972; Nelson 1977). In this work it was shown that the ratio of the dielectric loss of the insects to that of the wheat was greatest for radiation in this frequency range. Another example in which microwaves have been used with profit is the rapid warming of frozen organs and in particular foetal mouse hearts and canine kidneys (Voss, Rajotte, and Dossetor 1974).

The increased use of microwaves and lower frequency electromagnetic waves has brought with it potential hazards as well as benefits. The biological mechanisms which cause a steak to be cooked in a few minutes in a microwave oven could produce the same effects in the body of the chef if the source of the radiation were not adequately screened. In practice, commercial ovens are very well shielded, but the increasing use of them plus all the other uses of microwaves in industry, communications, and medicine does mean that exposure of the population to non-ionizing radiation is increasing, and therefore that the question of microwave hazards must be taken seriously. One form of biological damage which can be caused by microwaves of appropriate intensity is cataract. The lens of the eye is vulnerable to microwave injury partly because of its poor blood supply, and so fewer paths for heat to be conducted away; it is also significant that the lens has a higher proportion of water, both free and bound, which causes the energy to be readily dissipated in the first place. As well as the well established biological hazards of microwaves there are numerous reports of behavioural effects and other phenomena which fall in the realm of neurophysiology. There is still considerable disagreement between scientists in eastern and western European countries about the hazards of microwaves, and radically different safety levels are adopted. The proceedings of a fairly recent international symposium on microwave hazards have been published under the general editorship of Czerski (1974) and numerous conferences on the subject have been held over the past few years (e.g. Adey and Bawin 1977). The basic mechanisms of interaction have been discussed by Schwan (1977).

1.6. Dielectric studies as a tool in biophysics: résumé and outlook

It is obvious that the simpler the system being studied the greater the likelihood of obtaining precise information about it. In the present context this means that an aqueous solution containing only one kind of biological molecule as the solute is very amenable to study by dielectric methods. A particularly suitable system is an aqueous solution of a globular protein, for which reliable values of molecular shape, size, and hydration can be obtained (§ 6.3). In contrast with solutions of smaller biological molecules (molecular weight less than a few hundred) the molecular dipole moment (μ) is not given specifically from dielectric measurements; instead $k\mu$ is obtained, where k is a parameter whose value may be difficult to evaluate and which depends upon local solute—solvent interactions. This may make it difficult to obtain an accurate value of the dipole moment, but it does allow a comparison to be made between the dipole moments of molecules of similar size and structure. The presence or otherwise of dimers, trimers, and higher oligomers in solutions of small biological molecules can be usefully investigated by dielectric methods, and experience frequently shows that unimolecular rotation takes place and therefore that the formation of aggregates can be excluded. Attempts have been made to use dielectric methods to study bound water in solutions of small biological molecules. The difficulty with such investigations is that the three dispersion regions of the solute, the bound water, and the free water overlap. Therefore, we would not recommend this kind of study unless the investigator is prepared to measure permittivity and dielectric loss at many separate frequencies over a frequency range of at least four decades.

We must now consider the position of dielectric measurements in relation to the study of solutions of very large biological molecules, e.g. DNA or lipoproteins. Some of the bigger protein molecules which do not take up the form of a compact globular structure can also be included in this category. For all these molecules the polarization is not in general due to the rotation of the molecule as a rigid entity, and therefore precise values of molecular parameters cannot be obtained from a study of the dielectric parameters. However, useful information of another kind can be obtained. One example of this for elongated molecules is the detection of the change in the conformation of the molecule from a helix to a coil as a function of some physical parameter, e.g. temperature or pH. The large changes in dipole moment and relaxation time occurring during this helix-coil transition are strongly reflected in the permittivity measurements. Thus if an investigator considers that the phenomenon to be studied involves a change in dipole moment or confrontation, then the dielectric method could be used to follow this change.

With heterogeneous biological materials the amount of unambiguous information that can be obtained is inevitably limited by the complexity of the system. If a property is to be investigated which predominates over a particular frequency range, the background effect of the other components of the specimen may be small enough to enable useful measurements to be performed. For example, the characteristics of the membrane of the red blood cell can be studied by measurements made on whole blood at radio frequencies. It may be assumed that, of all the different types of molecule present in blood, only those of free water are able to rotate in the electric field in the very much higher frequency region of a few GHz. Therefore studies at frequencies higher than this will give information on how the water structure is affected by the other macromolecules present. Moreover, an estimate of the amount of bound water present may be obtained from a single determination of permittivity at a frequency near 1 GHz. Another use of measurements of the permittivity and conductivity of a complicated heterogeneous system is for the provision of qualitative information when a rigorous interpretation is not possible, for example, the establishment of the presence of charges over the surface of a virus or the variation of the electrical properties of tissues with metabolic status come to mind.

Obviously good quality equipment is necessary and in this connection it is important to remember that although most of this apparatus can be purchased commercially, the experimental cell has to be constructed in laboratory workshops. With bridge apparatus the construction of the experimental cell is somewhat easier than with coaxial lines, but careful machining is still required. When very precise bridge measurements are sought it may be reassuring to carry out experiments on two independent pieces of apparatus which have a range of frequency overlap.

The future progress of dielectric methods in biophysics will depend, on the experimental side, on the ability to devise methods which will measure *rapidly* the permittivity and conductivity of *conductive* solutions over a wide frequency range. Time domain techniques may be of considerable value in

this respect. The ability to be able to handle specimens of
conductivity higher than can at present be investigated would
open up a new range of materials that can be studied. Achievement of this will largely depend on the improvement in the
resolution which can be attained and, in the case of bridges,
on overcoming electrode effects (§ 3.2.2). On the theoretical
side, the power of the dielectric method will increase as
further understanding is gained of the local structure of the
region surrounding a biological molecule in solution, and as
advances are made in the general theory of liquids.

2
DIELECTRIC THEORY

This chapter is designed to give the uninitiated reader an introduction to the concepts of dielectrics and a brief account of the mathematical approach to general macroscopic theory and its relation to the molecular theory of dipolar rotation. It is by no means comprehensive, and for fuller and more specific descriptions the reader is referred to Hill (1969) and Scaife (1971). Reviews of the applications of the theory to systems other than biological ones are given by Davies (1965, 1969), Price (1969).

2.1. The nature of dielectrics
2.1.1. *Electric forces and fields*

A fundamental property of all dielectrics is the electric charge which arises directly out of the atomic structure of matter and in order to describe electrical phenomena, a unit in addition to those of mass, length, and time is required. It has now become conventional to define this unit as the 'ampere', the current which, when flowing along each of two infinitely long parallel and thin wires one metre apart, produces a force between the wires of 2×10^{-7} newtons per metre (Nm^{-1}) of their length. The 'coulomb' is then defined as the charge which flows in one second through a wire carrying a current of one ampere. Accurate measurement has shown that the charges of protons and electrons are both 1.602×10^{-19} coulomb (C), the electron being arbitrarily assigned a negative charge and the proton a positive one.

Much of the theory of electricity relies strongly on the 'inverse square law' of force between electric charges, which states that if two charges are separated by a distance r then the force which each feels as a result of the presence of the other is proportional to $1/r^2$. This is known as Coulomb's law and was tested experimentally by him using charged pith balls; more recent techniques have shown the law to be true to better than two parts in 10^9 (Plimpton and Lawton 1936). The force between two such charges is also proportional to the magnitude of the charges q_1 and q_2 and is generally written

$$F = \frac{q_1 q_2}{4\pi \varepsilon_0 r^2}, \qquad (2.1)$$

where ε_0 is a fundamental constant known as the permittivity of free space and has the value 8.854×10^{-12} in SI units (Fm^{-1}).

This inverse square law leads to the idea of lines of electric field. The electric field is a vector (i.e. a quantity having direction and magnitude) which can be assigned to each point in space and which in general varies from point to point. The magnitude of the vector is the force which a charge of one coulomb would feel if it were placed at the point in question; the direction of the electric field is the direction of this force. Alternatively, if a charge of q_2 is placed at a point where the electric field is E, then it will experience a force $q_2 E$. Comparing this with eqn (2.1) shows that the magnitude of the field at a distance of r from a point charge q_1 is

$$E_0 = \frac{q_1}{4\pi \varepsilon_0 r^2}. \qquad (2.2)$$

Electric field lines may now be introduced in such a way that their direction is that of the field and their density is proportional to the magnitude of the field. Thus for a point charge, the field lines are spokes pointing out symmetrically from the charge, as shown in Fig. 2.1. If a charge of $+q$ is

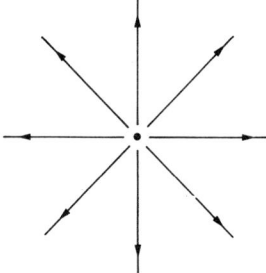

FIG. 2.1. Electric field lines radiate outwards from an isolated positive charge.

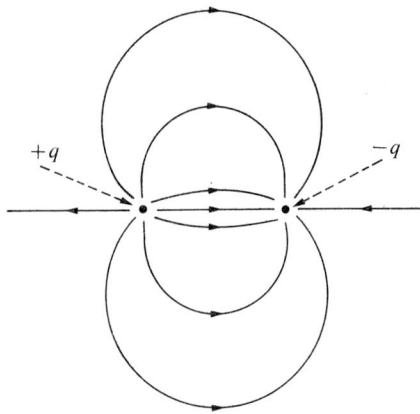

FIG. 2.2. Equal positive and negative charges displaced from each other constitute a dipole whose electric field geometry is shown here

placed at a distance from a charge of $-q$ then the field lines would be as shown in Fig. 2.2. The idea of field lines is possible because they spread out in three-dimensional space according to an inverse square law i.e. the same as the electrical forces. It may be seen in the figures that the electric field is different at each point, if not in magnitude then in direction, but if we consider a very small volume at some distance from the point charge then the electric field is nearly constant, as shown in Fig. 2.3. This field may be made as constant as we wish by taking the volume sufficiently small and distant from the charge. We will return to this small volume shortly.

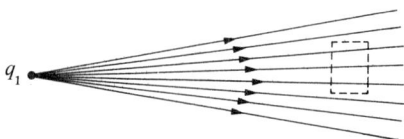

FIG. 2.3. Some of the field lines from a positive charge. A volume small compared with its distance from the charge experiences a nearly constant field.

2.1.2. Polarization

We now return to the two point charges separated by a distance r. Previously these were considered to be in an infinite vacuum, but if we now imagine this space to be filled with a homogeneous material, then it is found experimentally that the force between the two charges is reduced from its previous value and may be written

$$F = \frac{q_1 q_2}{4\pi \varepsilon_0 \varepsilon_s r^2}, \qquad (2.3)$$

where ε_s is a dimensionless constant having a value greater than unity; for oxygen it is 1.00051 and for water it is about 80 at room temperature. This number is the 'dielectric constant' or the 'relative permittivity' of the material and it is with this quantity and its frequency-dependent counterpart, $\hat{\varepsilon}$, that we are essentially concerned in this subject.

The definition of electric field in a material medium remains the same as before and hence the field, E_m, at a distance r from a point charge will now be given by

$$E_m = \frac{q_1}{4\pi \varepsilon_0 \varepsilon_s r^2}. \qquad (2.4)$$

There has thus been a reduction in the field of E_0 to E_0/ε_s and as electric field can only be caused by electric charges, there must be some charges which are producing the field $(1-1/\varepsilon_s)E_0$ in the opposite direction to the applied field, E_0. These charges arise from the 'polarization' of the medium in the following way. The field of q_1 tends to push the positive charges of the medium (i.e. the nuclei of the atoms) to the outer edge of the small volume in Fig. 2.3 and the negative charges (i.e. the electrons) to the inner edge; this separation of the charges is not complete owing to the forces between the nuclei and the electrons, and thus the charges seen by the small volume are now those shown in Fig. 2.4, where the induced charges produce a field in the opposite direction from the applied field.

This reduction in the field when a material medium is introduced is also seen in the case of the 'parallel plate

FIG. 2.4. The small volume of Fig. 2.3 experiences a field from the point charge and from the polarization charge when a material medium is present. These fields are in opposite directions

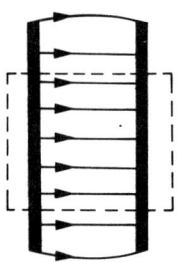

FIG. 2.5. A parallel plate capacitor with positive charge on its left side and an equal negative charge on the right side. The field lines are parallel within the dotted area

capacitor' shown in Fig. 2.5. This consists of two flat metal plates separated by a constant distance, d. One plate is given a positive charge and the other a negative charge of the same magnitude. These charges distribute themselves over the plates and produce field lines as shown in the figure. If only the volume within the dotted line is considered, then the field lines are straight and regularly spaced, i.e. the field is the same at all points. If the area of each plate within the dotted lines is A and the charges on the plates are $+q$ and $-q$, then the field may be shown to be

$$E_0 = \frac{q}{\epsilon_0 A} . \qquad (2.5)$$

If now the space between the plates is filled with a material of relative permittivity ϵ_s then again the field is reduced by a factor $1/\epsilon_s$ and becomes

DIELECTRIC THEORY

$$E_m = \frac{q}{\varepsilon_s \varepsilon_0 A}. \tag{2.6}$$

The field is that which would be produced if the charge on the plates were q/ε_s. This reduction in the apparent charge on the plates is caused by the polarization of the medium; in fact there is an induced charge throughout the material but it annuls itself everywhere except the surface. Thus the charge induced close to the left plate is $+(1-1/\varepsilon_s)q$ to give a total charge of $-q/\varepsilon_s$.

From eqn (2.6) the capacitance of the parallel plate capacitor may be found. This is defined as q/V where V is the difference of electric potential between the two plates. (Capacitance therefore has the units of CV^{-1} which is abbreviated to farads, F.) Since $V = E_m d$,

$$\frac{V}{d} = \frac{q}{\varepsilon_s \varepsilon_0 A};$$

whence $\quad C = q/V = \dfrac{\varepsilon_s \varepsilon_0 A}{d}. \tag{2.7}$

This relationship is particularly useful in the measurement of permittivity by bridge techniques (§ 3.2).

This electric polarization or 'electrical distortion' of the material may be likened to the physical distortion of a wire under tension. If the wire has a natural length l, and cross-sectional area S, and suffers an extension of e when subject to a tension T, then

$$e = Y \times \frac{l}{S} \times T \tag{2.8}$$

where Y is the Young's modulus of the material of the wire.

For electrical distortion the polarization charge, q_p, is

$$\begin{aligned} q_p &= q(1-1/\varepsilon_s) \\ &= (1-1/\varepsilon_s)(\varepsilon_0 \varepsilon_s A)\left(\frac{q}{\varepsilon_s \varepsilon_0 A}\right) \\ &= \varepsilon_0(\varepsilon_s-1)A\, E_m. \end{aligned} \tag{2.9}$$

Both eqns (2.8) and (2.9) have the form

distortion = constant of material × constant of geometry × 'tension'.

The induced charge, q_p, is not so suitable a quantity to treat theoretically or experimentally as is the 'dipole moment' which is defined, for the present system, as $q_p d$. This leads to the definition of dielectric polarization, P, as the induced dipole moment per unit volume i.e. $(q_p d)/(Ad)$, which is q_p/A. From eqn (2.9) this leads to

$$P = (\varepsilon_s - 1)\varepsilon_0 E_m, \tag{2.10}$$

thus removing the constant of geometry and supplying an equation which is not only true for a parallel plate capacitor but also at any point in a dielectric of any volume or shape. Since both the field and the polarization have direction as well as magnitude, they are vector quantities. In the cases of interest here (i.e. homogeneous dielectrics) these two vectors are parallel, but in some substances (e.g. crystals) this is not the case and ε_s is then a tensor.

We see, therefore, that dielectric permittivity is a measure of the response of a system to an applied electric field, in the same way that Young's modulus is a measure of physical distortion to an applied force. We look now for a more detailed molecular explanation of such polarization; there are many mechanisms by which it can occur and it is just this variety which gives dielectrics such a wide appeal to those interested in the structure of materials. Dielectric measurements are of value in dealing with the near perfect gases, simple liquids, solids such as alkali halides, long polymer chains and, as this book aims to show, water and biological material.

2.1.3. *Polar and non-polar dielectrics*

It has been usual to divide materials into those which are 'polar' and those which are non-polar. A polar dielectric is one in which the individual molecules possess a dipole moment even in the absence of any applied electric field, i.e. the centre of positive charge is displaced from the centre of

DIELECTRIC THEORY

negative charge. The value of the dipole moment of an uncharged molecule is given by total positive or negative charge multiplied by the separation of the two centres of charge and is generally given the symbol μ. The SI unit of dipole moment is that of a charge of +1 C at a distance of 1 m from a charge of -1 C. However, this leads to molecular dipole moments of the order of 10^{-30} C m, which is inconvenient for repeated use. For this reason the old Debye unit has been retained; this is the dipole of an electron and a proton separated by a distance of 2.38 nm; the conversion factor is 3.33×10^{-30} C m per Debye.

When a molecule having a permanent dipole moment is placed in an electric field, the positive end will tend to move in the direction of the field and the negative end in the opposite direction; the molecule will therefore tend to rotate as shown in Fig. 2.6. If there is no net charge on the molecule then it will experience no translational force. The molecule will not completely align itself in the field since it is also subject to the forces of Brownian motion transmitted by the random bombardment of its neighbours. In fact for the fields applied in typical dielectric work, the average angle of rotation caused by the field is only a small fraction of a degree, but for a system containing many such molecules the total effect is quite measurable and does give rise to high values of ε_s.

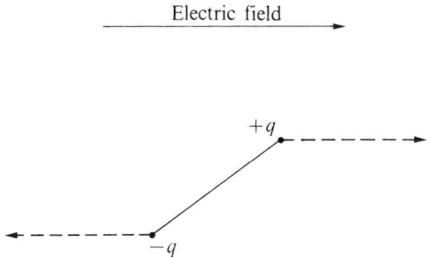

FIG. 2.6. A dipole experiences a couple in an electric field. The dotted lines represent the forces felt by each charge

The exact calculation of the dipole moment of a molecule from its molecular structure is an advanced exercise in quantum mechanics, but if this value is known then the calculation of the relative permittivity of an assembly of such molecules is largely a problem in classical statistical mechanics. The first solution of this problem was given by Debye (1929) in his well-known book *Polar molecules*.

A non-polar dielectric is one whose molecules possess no permanent dipole moment unless they are in the presence of an electric field. In this case the field induces molecular dipole moments by perturbing the electron cloud around the nucleus so as to produce a separation of the centre of negative charge and the centre of positive charge. This is shown in Fig. 2.7. In poly-electrolytes the counterion distribution is similarly perturbed by the field.

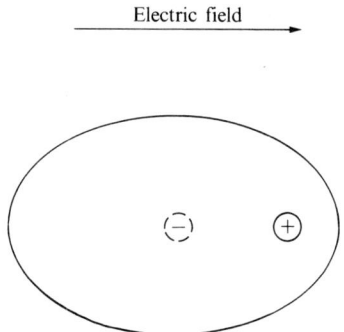

FIG. 2.7. A molecule becomes polarized in an electric field. The positive nucleus moves one way and electron orbitals the other. The dotted circle represents the effective centre of negative charge

The calculation of this effect is again a difficult problem in quantum mechanics and requires precise information on the molecular wavefunction. In practice this is rarely known and historically, therefore, dielectrics has been used to find values for these quantities by experiment, which have then been helpful in establishing the molecular structure.

The mechanisms of dipolar rotation and polarizability (which can also lead to molecular rotation) are the most common in considering the dielectric properties of simple substances and a great deal of information may be gained from these alone. In the field of biological dielectrics it is necessary to consider several other mechanisms, as will be seen later, but even then the concepts of molecular rotation and polarizability retain a central importance.

2.1.4. *Dielectric relaxation*

So far we have considered only the 'static' or 'equilibrium' situation, i.e. the system has reached a constant value of polarization, P, under the influence of a constant field, E. Much valuable information is also gained from the time dependence of dielectric processes. Considering again the analogy with the stretched wire, if the applied tension is suddenly removed the wire does not return to its natural length in zero time as this would require an infinite acceleration; the wire gradually returns to its natural length. The same is true of dielectric processes since it takes a finite time for the electric charges to 'relax' to their natural random positions and orientations. Thus if the field E is suddenly removed and P is measured, it would show a gradual decay towards zero, as shown in Fig. 2.8. The time scale and the exact shape of this decay are related to the structure of the material and the mechanism causing the polarization. If P results from the induced dipoles of the electron wavefunction, then it will decay in a very short time of the order of 10^{-14} s or less: if P is caused by the orientation of small molecules, the decay may take about 10^{-11} s (e.g. some liquids), whereas if P is caused by the orientation of large polar molecules (e.g. proteins in aqueous

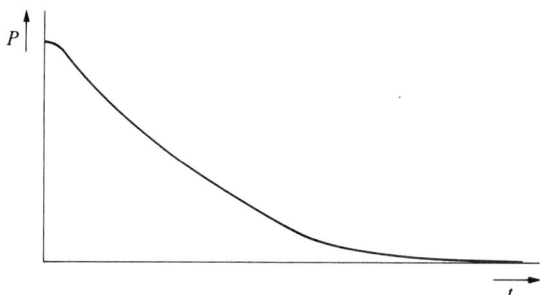

FIG. 2.8. Decay of polarization with time after a field has been removed

solution), the decay takes a time of order 10^{-8}–10^{-6} s; in some solid polymers the decay may take several seconds or even hours.

It may be asked what exactly is meant by the time of decay when the curve in Fig. 2.8 never seems to reach the zero axis but only to approach it asymptotically. A rigid definition will be given later, but for the moment the decay time may be taken as the period for the polarization to decay by a substantial fraction of its original value — say by one half. The information given by this decay time clearly depends on the mechanism which is causing the polarization, and in the case of molecular rotation in the Debye model it is related to the size and shape of the molecule together with the viscosity of its environment. Other possible interpretations of the decay rate, such as proton fluctuation times, will be discussed in later sections when they become relevant.

It is noticed in Fig. 2.8 that the slope of the decay curve tends to zero as t tends to zero; this is a necessary consequence of time reversibility and is brought about by inertial effects which are not of importance here. For clarity the effect has been exaggerated in the figure and may be neglected in the applications discussed in this book.

In summary, therefore, the relative permittivity provides information on how easily a system is distorted by an electric field and the decay of this distortion is related to the temporal properties of the system. In order to make further progress it is necessary to give these ideas a more mathematical foundation, and this will be done in the following sections of this chapter.

2.2. The Static Relative Permittivity
2.2.1. *The meaning of static permittivity*

The equation

$$P = (\varepsilon_s - 1)\varepsilon_0 E \tag{2.10}$$

relates the polarization per unit volume to the applied field. Firstly it should be realized that this is not a rigorous equation but only a first approximation, since in a real system there will be further terms on the right-hand side containing higher powers of E; these are the so-called 'non-linear' terms which, providing E is not too big, are quite negligible and will be of no further concern to us here. A further qualification to the formula is that it holds only for a homogeneous dielectric and is assumed to hold at every point in the system. This may give rise to some concern since we know that any real substance is composed of individual atoms and molecules, inside which the definition of P as given previously would be quite meaningless. To overcome this difficulty it is necessary that the system under consideration should be much larger than molecular dimensions, so that P may be taken as an average value over many molecules.

Although eqn (2.10) is an equilibrium equation, this does not mean that the system is not changing with time; the individual molecules are in a rapid state of translation and rotation owing to Brownian motion. The equilibrium permittivity is therefore the result of a dynamic equilibrium in the system. If P was carefully measured it would be seen to be fluctuating randomly about its mean value because the individual molecular dipole moments would sum to slightly different values at two different instants of time. However, because there are many such dipoles fluctuating randomly with respect to each other, the sum never deviates much from the mean value and thus the fluctuation in P is just a small ripple.

It may perhaps be thought that such Brownian motion effects are an unfortunate complication, but it should be remembered that were it not for these, the relative permittivity would be meaningless. This is most easily seen in the case of dipolar rotation where all the permanent dipole moments would completely align themselves with even the smallest of applied fields and the permittivity could take almost any value. However, in reality the Brownian motion forces tend to randomize this alignment and eventually a dynamic equilibrium is set up in which the molecules are being aligned by the field at the same rate as they are being randomized by the Brownian motion. The relative permittivity is just a measure of the competition between these two effects in a similar way to the dissassociation constant being a measure of the position at which a chemical reaction reaches dynamic equilibrium.

The close relation between fluctuation effects and equilibrium permittivity has been used extensively in some recent developments in dielectric theory which will be discussed later. For the time being, we are interested in just the mean value of P when a constant field, E, is applied to the system, and it is this value that must be used in eqn (2.10).

2.2.2. *The electric displacement vector*

We now introduce a new vector quantity, the displacement **D**, which is defined as

$$\mathbf{D} = \varepsilon_0 \mathbf{E} + \mathbf{P}. \tag{2.11}$$

The less mathematical reader may wish to omit this part of the development of dielectric theory since, although it gives a more rigorous foundation to the idea of relative permittivity, it is not essential to an understanding of molecular interpretation of dielectrics.

In a vacuum the electric field may be found at any point from the Poisson source equation (Bleaney and Bleaney 1976):

$$\nabla \cdot \mathbf{E} = \rho/\varepsilon_0, \tag{2.12}$$

where ρ is the charge density and ∇ is the vector operator $\mathbf{i}\, \partial/\partial x + \mathbf{j}\, \partial/\partial y + \mathbf{k}\, \partial/\partial z$, where $\mathbf{i}, \mathbf{j}, \mathbf{k}$ are the unit vectors along the three perpendicular Cartesian axes.

In a material medium the charge density is not only that of the free charges, since there is also a contribution from the non-uniform distribution of charge induced in the medium by the applied field. Hence

$$\nabla \cdot \mathbf{E} = \frac{(\rho + \varepsilon_p)}{\varepsilon_0}, \tag{2.13}$$

where ρ is the free charge density and ρ_p is the polarization charge. Thus

DIELECTRIC THEORY 39

$$\nabla \cdot (\mathbf{E} + \mathbf{p}/\varepsilon_0) = \rho/\varepsilon_0$$

$$\nabla \cdot \mathbf{D} = \rho, \qquad (2.14)$$

where from eqn (2.11)

$$\mathbf{D} = \varepsilon_0 \mathbf{E} + \mathbf{P}.$$

Thus **D** is a vector field which can be derived from the free charges only and it is this which makes it a useful quantity. It also follows from eqn (2.10) that

$$\mathbf{D} = \varepsilon_s \varepsilon_0 \mathbf{E} \qquad (2.15)$$

and therefore the permittivity is a constant of proportionality between **D** and **E** in any medium. A high value of ε_s implies a large induced charge density.

2.2.3. *Theory of the static permittivity of fluids*

Knowledge of the value of ε_s is of little interpretive value unless a theory is available which relates it to structural parameters of the system in question. Since this structure may take a wide variety of forms, it is helpful if the theory will go as far as possible before any specific model is introduced. One useful and neat means of achieving this is the use of the fluctuations of the system.

Consider a situation in which a constant vector **F** (which may be a force, electric field, magnetic field etc.) acts on an isotropic system and produces a resultant effect **R** (physical distortion, polarization, magnetization etc.). If, as is often the case, the interaction energy of **F** with **R** is -**F.R**, then it may be shown that the 'static susceptibility', α_s, of the process is given by

$$\alpha_s = \frac{R}{F} = \frac{\langle R^2 \rangle_0}{3kT}, \qquad (2.16)$$

where $\langle R^2 \rangle_0$ represents the mean square fluctuations of the quantity **R** in the absence of the applied force **F**, as is shown in Fig. 2.9. In this figure just one of the three components

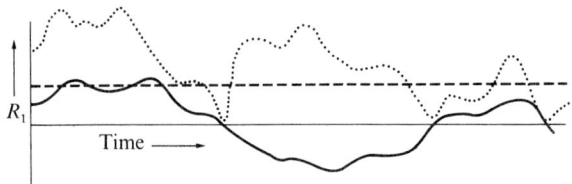

FIG. 2.9. Random fluctuations. Solid curve: magnitude of fluctuation R_1. Dotted curve: R_1^2. Dashed curve: mean value of R_1^2 ($\langle R_1^2 \rangle$)

of R is shown as a random fluctuation in time; although the mean value ($\langle R_1 \rangle_0$) is zero, it is evident that the mean square value is finite. Further detail of this approach to fluctuations is given by Kittel (1971).

It is seen, therefore, that the information given by the susceptibility is also manifest in the system when it is experiencing no applied force. For instance, the amount by which a wire will stretch under a given tension could be predicted by studying the small but finite fluctuations in its length when under no tension. It will be seen in § 2.3.4 that the frequency dependent susceptibility is also contained within these natural fluctuations of the system.

In the case of dielectric polarization, **R** represents the dipole moment of the system which is non-zero even in the absence of a field owing to the spontaneous fluctuations of electric charge which occur in a random way as a result of thermal energy. If the sample were in the form of a large thin slab, the susceptibility (dipole moment/applied field) could be simply related to ($\varepsilon_s - 1$); however, it is more convenient theoretically to consider a spherical sample which is immersed either in a vacuum or in an infinite extent of its own medium. In the latter case, which will be considered here, it may be shown that (Fröhlich 1958; Hill 1972)

$$\alpha_s = \frac{(\varepsilon_s - 1)(2\varepsilon_s + 1)}{3\varepsilon_s} \varepsilon_0 \, v \,, \qquad (2.17)$$

where the applied field is that which would be present in the spherical cavity if the material were removed. Hence from eqn (2.16) it follows that

$$\frac{(\varepsilon_s-1)(2\varepsilon_s+1)}{\varepsilon_s} = \frac{\langle M_T^2 \rangle_0}{\varepsilon_0 \, vkT} \qquad (2.18)$$

where $\langle M_T^2 \rangle_0$ represents the mean square fluctuation of the sphere. This general equation applies to any linear system be it solid, liquid, or gas and for whatever mechanism creates the polarization. To proceed further some model must be used to evaluate $\langle M_T^2 \rangle_0$ and hence to relate to molecular quantities.

A simple model appropriate for a fluid in which there are no local forces between molecules and no induced dipoles is that of a set of randomly rotating permanent dipoles μ_i ($i = 1, 2, \ldots, N$). In this case

$$M_T = \sum_{i=1}^{N} \mu_i \, . \qquad (2.19)$$

Thus

$$\langle M_T^2 \rangle_0 = N \langle \mu_i^2 \rangle = N\mu^2, \qquad (2.20)$$

since cross-correlation terms vanish and μ is the magnitude of the molecular dipole moment. Hence ε_s is given by

$$\frac{(\varepsilon_s-1)(2\varepsilon_s+1)}{\varepsilon_s} = \frac{N\mu^2}{\varepsilon_0 \, vkT} \, . \qquad (2.21)$$

A model which is better suited to many pure liquids is that of Fröhlich (1958) in which molecules possessing a permanent dipole moment μ are dispersed in a medium of permittivity ε_∞ which represents the induced polarization of the molecules. Following through a similar development to that above yields the equation

$$\frac{(\varepsilon_s-\varepsilon_\infty)(2\varepsilon_s+\varepsilon_\infty)}{\varepsilon_s} = \frac{Ng\mu^2}{\varepsilon_0 \, vkT} \, . \qquad (2.22)$$

The quantity g (Kirkwood correlation factor) accounts for the fact that short-range interactions between molecules may lead to a net polarization in the immediate environment of a given

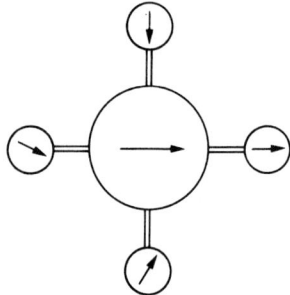

FIG. 2.10. Chemical forces orientate neighbouring molecules. In this case this enhances the dipole of the central molecule and g, the Kirkwood correlation parameter, would be greater than unity.

molecule (Fig. 2.10). The parameter g may be defined by equating $g\mu$ to the vector sum of the central molecular dipole and the dipoles of its neighbours interacting with it with short-range forces. Hence in principle g is a tensor, but in the application of interest here it is sufficient to describe it by a scalar which in many cases differs only slightly from unity. If this increases the dipole moment in this region then $g > 1$ and vice versa. The dipole μ is not that of the isolated molecule, i.e. that which would be measured in the gas phase, μ_g, since the molecule is polarized by the field of this dipole; it can readily be shown that

$$\mu_g = \frac{3}{\varepsilon_\infty + 2} \mu \ . \qquad (2.23)$$

When this is combined with eqn (2.22) Fröhlich's formula results,

$$\frac{(\varepsilon_s - \varepsilon_\infty)(2\varepsilon_s + \varepsilon_\infty)}{\varepsilon_s (\varepsilon_\infty + 2)^2} = \frac{N g \mu_g^2}{9kTv\varepsilon_0} \ . \qquad (2.24)$$

This has as special cases the older formulae of Kirkwood (1932, 1939) and Onsager (1936) when $\varepsilon_\infty = 1$ and $g = 1$ respectively.

Of particular interest to biological applications is the case of a mixture of polar molecules; if these are distinguished by subscripts A and B then

DIELECTRIC THEORY

$$\langle M^2 \rangle = g_A N_A \mu_A^2 + g_B N_B \mu_B^2 \tag{2.25}$$

and it follows that

$$\frac{(\varepsilon_s - \varepsilon_\infty)(2\varepsilon_s + \varepsilon_\infty)}{\varepsilon_s} = \frac{1}{kT v \varepsilon_0} \left\{ g_A N_A \mu_A^2 + g_B N_B \mu_B^2 \right\}, \tag{2.26}$$

where μ_A and μ_B represent the dipole moments created by the fixed charges in the molecule.

The two applications of eqn (2.18) which have been given here are concerned with dipolar rotation as the mechanism creating a mean square dipole moment; nevertheless, eqn (2.18) is valid whatever the mechanism.

This fluctuation in the total dipole moment of a system has also been used to measure permittivity (§ 3.5).

2.3. Time Dependence and the Complex Permittivity

2.3.1. *Relaxation and resonance*

There are two distinct types of time dependence which occur in nature. The first of these is the regular oscillation of a system or a part of a system at a definite frequency, e.g. intra-molecular vibrations. Such oscillations will absorb energy from a suitable input signal over a narrow range of frequencies close to the resonant frequency. The second type is a far more general and random fluctuation which is caused by the thermal energy, $\tfrac{1}{2}kT$, associated with each degree of freedom of a system. These random fluctuations absorb energy in significant amounts if the frequency of the input signal (assumed to be sinusoidal) is sufficiently high, if the frequency is increased further then the absorption of energy is not reduced. These two situations are shown in Fig. 2.11.

The study of the resonance type of time dependence usually comes under the heading of 'spectroscopy', whereas the relaxation type (in the electrical case) is known as dielectric relaxation. These two have usually been treated separately since resonance generally occurs at higher frequencies than relaxation, and hence different types of apparatus have been

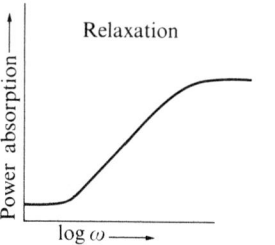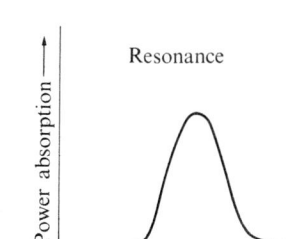

FIG. 2.11. Difference between power absorption in relaxation and in resonance

used in their study. Recent advances in experimental technique in the form of reliable apparatus operating at tens of GHz (§ 3.4) are bringing these subjects together, although they often remain separated on theoretical grounds in that resonance is essentially a topic of quantum mechanics whereas relaxation can generally be described in terms of classical statistical mechanics. It is with the latter that we are concerned here.

When the orienting (or more generally the ordering) force is removed, only the randomizing forces remain and it is the nature of these forces which determines the time necessary for the system to become randomized. In the language of § 2.2.2 this is the time necessary for the induced charge density, ρ_p, to decay.

2.3.2. *Dielectric relaxation and complex permittivity*

In practice it has become usual to measure the relaxation not directly, but by finding the relative permittivity of the system at different frequencies of applied sinusoidal field. (However, see § 3.6.) If the applied frequency is very low then the displacement or polarization at any time will be that appropriate for the value of electric field which obtains at that time, i.e. from eqn (2.10)

$$P(t) = (\varepsilon_s - 1)\varepsilon_0 E(t). \qquad (2.27)$$

If the frequency of the applied field is increased, the point will come when the polarization of the system is no longer able to follow the more rapid changes of the electric field. On the molecular orientation model, this corresponds to the field changing at a faster rate than that at which the molecular dipoles are able to rotate.

Eqn (2.27) may still be used provided that ε_s is changed to the complex number

$$\hat{\varepsilon} = \varepsilon' - i\varepsilon'', \tag{2.28}$$

where $i^2 = -1$, ε' is the real part, and ε'' is the imaginary part of the complex permittivity.

This may be seen as follows. If the applied field is $E \sin(\omega t)$ where E is the magnitude of the field and $\omega (= 2\pi f)$ is the angular frequency, then using Im to stand for the 'imaginary part of',

$$E(t) = Im\ E\ e^{i\omega t} \tag{2.29}$$

since $e^{i\omega t} = \cos(\omega t) + i \sin(\omega t)$; eqn (2.15) then becomes

$$\begin{aligned} D(t) &= \varepsilon_0\ Im\ E(\varepsilon' - i\varepsilon'')e^{i\omega t} \\ &= \varepsilon_0\ Im\ E(\varepsilon'^2 + \varepsilon''^2)^{\frac{1}{2}}\ e^{i(\omega t - \varphi)}, \end{aligned} \tag{2.30}$$

where $\tan \varphi = \varepsilon''/\varepsilon'$.

Thus a complex permittivity (i.e. a finite ε'') implies a phase lag of φ between **D** and **E**. It is therefore apparent that as the frequency of the applied field is raised towards the relaxation region of the system, the value of φ ceases to be zero and will take increasing positive values. The presence of this phase difference between **D** and **E** causes an important absorption of energy by the system. If ε'' is zero, then the energy required to build up the electric field during one half of the cycle is exactly recuperated during the next half of the cycle and there is no net absorption of energy. However, if there is any phase difference between **D** and **E** then this is not the case and there is a net absorption of energy. This may be derived from Maxwell's equations of electromagnetism (Maxwell 1892; Bleaney and Bleaney 1976), which show that the presence of a finite ε'' has the effect of producing an electrical conductivity of magnitude $\varepsilon''\varepsilon_0\omega$, and that a finite electrical resistance, e.g. the element of an electric fire, always absorbs energy from a current flowing through it.

Some confusion can occur as a result of conductivity (or energy loss) arising from either dielectric relaxation or from more common conductive processes as in a metal or ionic solution. In biological systems both effects are generally present and, as a result of the mathematical identities mentioned above, it is impossible to separate the two contributions from measurements at one isolated frequency. Here we refer to the total frequency-dependent conductivity as σ (units $\Omega^{-1}\,m^{-1}$) and the total imaginary part of $\hat{\varepsilon}$ as ε''; the contributions from dielectric relaxation alone are σ_D and ε''_D and the conductivity at zero frequency (i.e. that arising from ionic mobility) is σ_s. The following general relationships therefore exist:

$$\sigma = \sigma_s + \sigma_D \tag{2.32}$$

$$\varepsilon'' = \sigma/\varepsilon_0\omega \tag{2.33}$$

$$\varepsilon''_D = \sigma_D/\varepsilon_0\omega . \tag{2.34}$$

For several reasons which are apparent in later sections of this book, it is useful to divide conductivity values by ε_0, thus giving them units of s^{-1}. These values are thus defined by

$$s = \sigma/\varepsilon_0 \tag{2.35}$$

$$s_D = \sigma_D/\varepsilon_0, \tag{2.36}$$

from which it follows that

$$s = s_s + s_D \tag{2.37}$$

$$\varepsilon'' = s/\omega \tag{2.38}$$

$$\varepsilon''_D = s_D/\omega . \tag{2.39}$$

DIELECTRIC THEORY

These quantities are shown graphically for the special case of a single relaxation time (see § 2.3.3) in Fig. 2.12.

2.3.3. *Dispersion equations*

It is clear from the above that ε' and ε'' are functions of frequency, ω, ε_s being merely the value of ε' when $\omega = 0$ or, in practice, at frequencies very much lower than these where dispersion occurs. The exact nature of these functions depends on the way in which a polarization $P(0)$ decays when the steady field producing it is suddenly removed at time $t = 0$; if the polarization at some subsequent time t is $P(t)$, then in general

$$\Phi(t) = \frac{P(t)}{P(0)}, \qquad (2.40)$$

where $\Phi(t)$ is the 'decay function' of the system and is clearly unity when $t = 0$ and tends to zero as $t \to \infty$. The frequency dependence of ε' and ε'' may then be evaluated by use of a very general expression known as the superposition theorem. This is a mathematical technique corresponding to the application of an infinite number of pulses to the system, which then responds to each one according to the time dependence of eqn (2.40); the pulses add up to a continuous sinusoidal signal and the responses are superimposed on each other to give the total response of the system (Hill 1969). Mathematically this may be described using the generalized quantities mentioned in § 2.2.3. If R has a decay function $\Phi(t)$ then the susceptibility α when the field

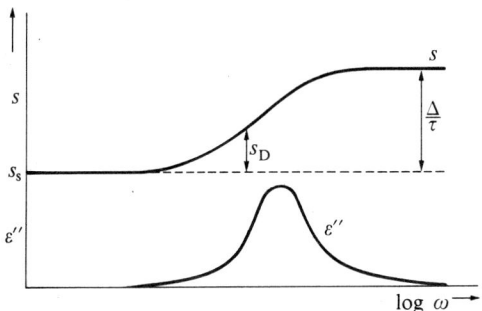

FIG. 2.12. Shape of ε'' and s dispersion curves for a single relaxation time

is varied sinusoidally with angular frequency ω is

$$\hat{\alpha} = \alpha_s \int_0^\infty \frac{d\Phi}{dt} e^{-i\omega t} \, dt \, . \tag{2.41}$$

The most simple specific case to consider is that which arises when $\Phi(t) = \exp(-t/\tau)$ where τ is the relaxation time of the system. Electric polarization rarely has exactly this decay function but in practice it is often a good approximation; it corresponds to the situation in which the rate of decay of $P(t)$ to zero is proportional to $P(t)$ itself i.e.

$$\frac{dP(t)}{dt} = - \frac{P(t)}{\tau} \, . \tag{2.42}$$

In this case integration of eqn (2.41) leads to

$$\hat{\varepsilon} - \varepsilon_\infty = \frac{\Delta}{1 + i\omega\tau} \tag{2.43}$$

or

$$\varepsilon' - \varepsilon_\infty = \frac{\Delta}{1 + \omega^2\tau^2} \tag{2.44}$$

$$\varepsilon''_D = \frac{\Delta\omega\tau}{1 + \omega^2\tau^2} \, , \tag{2.45}$$

where Δ and ε_∞ are the constants indicated in Fig. 2.13. These expressions are commonly known as the Debye dispersion equations

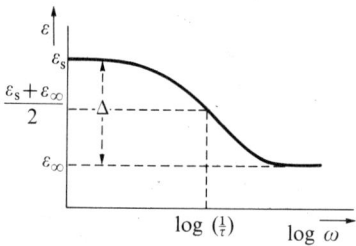

FIG. 2.13. The Debye dispersion: ε' as a function of log (frequency)

DIELECTRIC THEORY

and are frequently encountered in dielectric literature.

In terms of conductivity, eqn (2.45) becomes

$$s = s_s + \frac{\Delta}{\tau} \frac{\omega^2 \tau^2}{1 + \omega^2 \tau^2} . \qquad (2.46)$$

This relationship is shown in Fig. 2.12.

It is rare for a system to conform exactly to these equations and more usually the dispersion curve is found to be rather broader than that in Figs. 2.12 and 2.13. This may be explained if the system has more than one relaxation time and the decay of polarization is given by several terms with different relaxation times:

$$\mathbf{P}(t) = \sum_i P_i\, e^{-t/\tau_i} , \qquad (2.47)$$

in which case

$$\varepsilon' = \sum_i \frac{\Delta_i}{1 + \omega^2 \tau_i^2} + \varepsilon_\infty \qquad (2.48)$$

and

$$\varepsilon_D'' = \sum_i \frac{\Delta_i \omega \tau_i}{1 + \omega^2 \tau_i^2} . \qquad (2.49)$$

Sometimes this series of terms may consist of many relaxation regions close together, in which case an integral may be used:

$$\mathbf{P}(t) = P_0 \int_0^\infty G(\tau)\, e^{-t/\tau}\, d\tau. \qquad (2.50)$$

Hence

$$\varepsilon' = \Delta \int_0^\infty \frac{G(\tau)\, d\tau}{1 + \omega^2 \tau^2} + \varepsilon_\infty \qquad (2.51)$$

and

$$\varepsilon_D'' = \Delta \int_0^\infty \frac{G(\tau)\omega\tau\, d\tau}{1 + \omega^2 \tau^2} . \qquad (2.52)$$

Further consideration of this approach to the analysis of dielectric data is given in § 4.4.

It is interesting to note that whichever expression is best able to describe the dispersion curve, many systems show a variation of relaxation time (or times) with temperature described by the expression

$$\log_e(\tau T) = \frac{A}{T} + B \qquad (2.53)$$

where A and B are constants. This equation has a theoretical foundation in the model of an activation process, familiar in chemical reaction theory and described by Eyring, Glasstone, and Laidler (1941). This suggests that the relaxation time is

$$\tau = \frac{h}{kT} e^{\Delta F/RT}, \qquad (2.54)$$

where h is Planck's constant and ΔF is the difference between the free energy of the activated state and the ground state. This can be split into entropy (ΔS) and enthalpy (ΔH) contributions, giving

$$\tau = \frac{h}{kT} e^{-\Delta S/R} e^{\Delta H/RT}, \qquad (2.55)$$

which after taking logs shows that the constant, A, in eqn (2.53) may be identified with $\Delta H/R$. Calculation of ΔH is useful when comparing the molecular mechanisms underlying dispersion curves; for instance it is possible that although two relaxation times may be of very different magnitudes, their activation enthalpies are the same, implying some connection between the two processes. An example of this is the relaxation of water and protein in an aqueous protein solution; at room temperature water has a relaxation time around 10^{-11} s and small protein around 5×10^{-8} s, but both show activation enthalpies close to 17 kJ/mol.

2.3.4. *Theory of frequency-dependent permittivity*

The macroscopic theory outlined above must be related to the microscopic structure of the system for the purposes of interpretation. As for the static permittivity, it is convenient to develop the formalism as far as possible before the introduction of any specific model, and consideration of the natural fluctuations in dipole moment is again useful. The two fluctuations shown in Fig. 2.14, for instance, may very well have the same value of $\langle M^2 \rangle_0$, but they nevertheless have their own characteristic behaviour which is evident in the rapidity of the fluctuations. This is given a more rigorous description by the correlation function of the process, defined as

$$\Phi(t) = \frac{\langle M(0) \cdot M(t) \rangle_0}{\langle M^2(0) \rangle_0}. \quad (2.56)$$

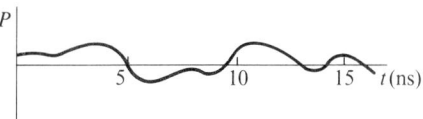

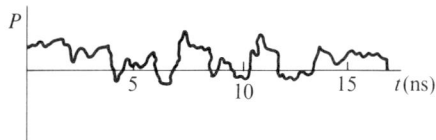

FIG. 2.14. Random fluctuations in polarization for two different systems

The behaviour of this function may be seen in the following way. If t is so small that very little change takes place in this interval, then

$$M(t) \approx M(0), \quad (2.57)$$

and hence the correlation function is close to unity. If, howevery, t is so large that considerable fluctuation takes place, then there will be little correlation between $M(t)$ and $M(0)$ and the correlation function approaches zero as $t \to \infty$.

This correlation function contains all the temporal information of the system and is therefore related to the frequency dependence of the permittivity. Stated in this way the situation is an example of a general theory of irreversible processes constructed by Kubo (1957). Several authors have made use of this in relating the frequency dependent permittivity to molecular rotation rates (Glarum 1960; Cole 1965; Fatuzzo and Mason 1967). Hill (1972) has pointed out that if it is assumed that natural fluctuations decay in the same way as larger induced perturbations (Onsager 1931), then the correlation function defined above is identical to the decay function described in the previous section. This link enables the superposition theorem to be used to derive a simple result, namely

$$\frac{(\hat{\varepsilon}-1)(2\hat{\varepsilon}+1)}{\hat{\varepsilon}} = \frac{(\varepsilon_s-1)(2\varepsilon_s+1)}{\varepsilon_s} L\{-\dot{\Phi}\}, \qquad (2.58)$$

where $\Phi(t)$ is the correlation function of the immersed sphere considered previously, and the following abbreviations are used:

$$\dot{\Phi} = \frac{d\Phi(t)}{dt} \qquad (2.59)$$

and

$$L\{f(t)\} = \int_0^\infty f(t) e^{-i\omega t} \, dt . \qquad (2.60)$$

This is the Laplace transform of $f(t)$, details of which are given in most texts on mathematical physics, e.g. Chisholm and Morris (1966).

Eqn (2.58) is quite general and holds for whatever mechanism(s) is creating the dispersion. The simplest application of this formula is to the system of randomly rotating point dipoles for which ε_s is given by eqn (2.21). In this case

$$\mathbf{M}(t) = \sum_{i=1}^N \boldsymbol{\mu}_i(t) \qquad (2.61)$$

$$\langle \mathbf{M}(0) \cdot \mathbf{M}(t) \rangle_0 = \sum_{i=1}^N \langle \boldsymbol{\mu}_i(0) \cdot \boldsymbol{\mu}_i(t) \rangle_0 \qquad (2.62)$$

since cross-correlation terms cancel. Hence

$$\Phi(t) = \frac{N\langle \mu(0) \cdot \mu(t) \rangle_0}{N\langle \mu^2(0) \rangle_0}$$

$$= \varphi(t), \qquad (2.63)$$

where $\varphi(t)$ is the microscopic or molecular correlation function, i.e. it refers to a single molecule rather than the whole system. If an ensemble of aligned molecules is imagined at $t = 0$ (Fig. 2.15(a)) and each molecule is then permitted to carry out Brownian rotation, then a little later the average alignment will have decreased (Fig. 2.15(b)) and will tend to zero as $t \to \infty$; this decay is described by $\varphi(t)$. For simple Brownian motion $\varphi(t) = \exp(-t/\tau)$ and hence

$$\frac{(\hat{\varepsilon}-1)(2\hat{\varepsilon}+1)}{\hat{\varepsilon}} = \frac{(\varepsilon_s-1)(2\varepsilon_s+1)}{\varepsilon_s} \frac{1}{1+i\omega\tau}, \qquad (2.64)$$

which gives the dispersion equation for $\hat{\varepsilon}$.

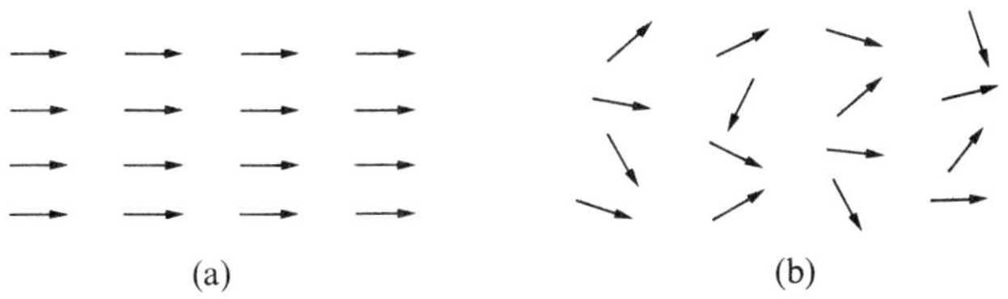

FIG. 2.15. (a) An ensemble of aligned dipoles. (b) The ensemble of dipoles after it has experienced Brownian rotation

For more complicated models the evaluation of $\Phi(t)$ becomes more difficult and the inclusion of induced polarization in particular leads to problems. Hill (1972) concludes that the best expression for taking this into account is

$$\frac{(\hat{\varepsilon}-\varepsilon_\infty)(2\hat{\varepsilon}+1)}{\hat{\varepsilon}} = \frac{(\varepsilon_s-\varepsilon_\infty)(2\varepsilon_s+1)}{\varepsilon_s} L\{-\dot{\varphi}_{\text{dip}}(t)\}, \qquad (2.65)$$

where φ_{dip} is the correlation function of that part of the polarization which may be attributed to permanent dipoles only.

A parallel approach may be applied to the case of the mixture of polar molecules whose static permittivity was given by eqn (2.26), resulting in

$$\frac{(\hat{\varepsilon}-\varepsilon_\infty)(2\hat{\varepsilon}+\varepsilon_\infty)}{\hat{\varepsilon}} = \frac{1}{kTV\varepsilon_0}\left[g_A N_A \mu_A^2 L\{-\dot\phi_A\} + g_B N_B \mu_B^2 L\{-\dot\phi_B\}\right]. \quad (2.66)$$

Of particular interest is the case of a solute molecule (A) considerably larger than the solvent (B). Two well separated dispersions occur, as shown in Fig. 2.16. At intermediate frequencies $L\{-\dot\phi_A\} = 0$ and $L\{-\dot\phi_B\} = 1$, hence if $\varepsilon_s \gg \varepsilon_\infty$ (as it is for a protein solution) then the dielectric increment, Δ, caused by the solute molecules is given by

$$\Delta = \frac{N_A g_A \mu_A^2}{2kTV\varepsilon_0}, \quad (2.67)$$

which, as seen in Chapters 5 and 6, is of considerable value in calculating dipole moments of biological molecules.

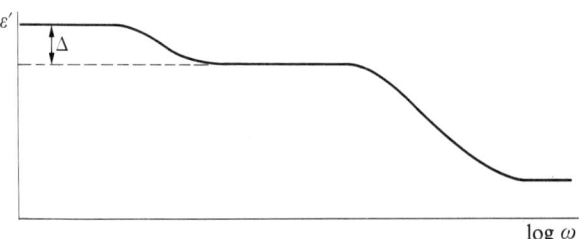

FIG. 2.16. Dispersion curve of a system with two widely differing relaxation times, e.g. a macromolecular solution. Δ is the increment of the lower frequency dispersion.

Now that the permittivity of two models has been expressed in molecular parameters, it is noticeable that some inconsistency exists with the previously described Debye dispersion curves; for example, eqns (2.43) and (2.64) do not give the same form for $\hat{\varepsilon}$. Eqn (2.43) was derived from the assumption that the decay of macroscopic polarization was exponential, but in eqn (2.64) it was the molecular correlation function which was given an exponential form. Thus the effect of electrostatic interactions between molecules is to transform an exponential molecular correlation function into a non-exponential macroscopic correlation function. In general, this has some important consequences, but in most biological work it is not important since very often $|\hat{\varepsilon}| \gg 1$ and hence expressions such as eqn (2.64) do approximate well to the Debye form.

Although in simple models of Brownian rotation the molecular correlation does have an exponential form, this may not be the case in practice. For instance a molecule of spheroidal shape subject to simple Brownian rotation gives the sum of two exponential terms. Furthermore, inter-molecular interactions may disturb the simple Brownian rotation and the molecular correlation function may be modified in such a way that it may be described by an integral summation of exponential terms:

$$\varphi(t) = \int_0^\infty g(\tau) e^{-t/\tau} \, d\tau, \qquad (2.68)$$

where $g(\tau)$ is a function which has a maximum at some particular value of τ and tends to zero on either side of this value. If the electrostatic factor described above is negligible then this leads to a dispersion curve of the type given in eqn (2.51). Many empirical expressions have been suggested for $g(\tau)$ and often these have been given in terms of the variable $\lambda = \log(\tau)$, i.e.

$$h(\lambda) d\lambda = g(\tau) d\tau . \qquad (2.69)$$

It was mentioned earlier that a model of relaxation based on rate processes leads to a direct relationship between λ and ΔH (eqn (2.55)) and hence a distribution of values of ΔH gives a similar distribution of λ. Fröhlich (1958) suggested the rectangular distribution shown in Fig. 2.17, but it is physically more reasonable to smooth this function into the Gaussian distribution which is shown superimposed on it; this is given by

$$h(\lambda) = \frac{1}{\sqrt{2\pi}\,\sigma_G} e^{-\frac{1}{2}\left(\frac{\lambda-\lambda_0}{\sigma_G}\right)^2}, \qquad (2.70)$$

where the centre of the distribution is given by λ_0 and σ_G is the standard deviation which is a measure of the width.

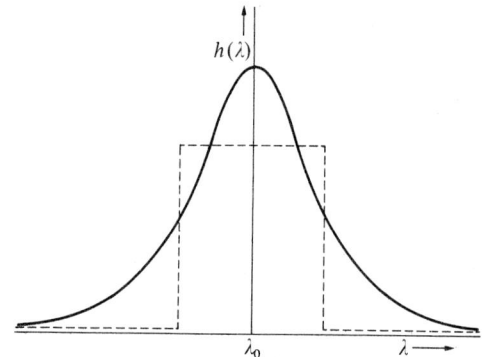

FIG. 2.17. Solid curve: Gaussian distribution. Dashed curve: Fröhlich rectangular distribution. (See text for explanation of symbols.)

These considerations have been based on dipolar rotation as the mechanism of dispersion, but it must be emphasized that eqn (2.58) is valid for any mechanism and hence the analysis of frequency-dependent permittivity must be related to some relevant model; this is considered in Chapter 4.

3
THE MEASUREMENT OF PERMITTIVITY

3.1. Introduction

This chapter describes experimental techniques which enable the complex permittivity of biological solutions to be determined. Since relaxation, rather than resonance, processes are likely to be encountered the frequency range of the dielectric dispersion generally covers at least two orders of magnitude, and for a heterogeneous system will be much greater than this. In order to characterize a range of biological substances by dielectric methods, apparatus able to cover a very broad frequency band is required. To be more specific, consider the dispersion curve for muscle tissue (Fig. 1.2) which ranges from below 100 Hz to 100 GHz. However, in many practical investigations much useful information can be obtained from a rather more restricted frequency coverage.

It must be stressed that this chapter will not attempt to survey dielectric measuring techniques generally, but will concentrate exclusively upon those methods most suited to biological solutions. Such solutions will in general be characterized by a high permittivity and a high conductivity.

The traditional methods of determining permittivity are centred around measurements in the frequency domain, the common feature of all such methods being that the solution is contained in a sample holder and its complex permittivity measured at various discrete frequency points. It is, however, not possible to devise one piece of apparatus or even one technique which can give the required frequency coverage.

At the low-frequency end of the spectrum, bridge techniques are used. Although these can in theory extend down to d.c., practical measurements on conductive biological solutions are difficult below about 1 kHz because of electrode effects (§ 3.2.2). The upper frequency limit of bridges tends to be between 100 and 250 MHz. Such methods are described in § 3.2.

At higher frequencies transmission lines are employed; such techniques are subdivided into coaxial lines and waveguides.

The frequency range of a coaxial line in practice is from around 50 MHz to possibly 12 GHz; above this and below 100 GHz waveguides are employed. These techniques are described in §§ 3.3 and 3.4 respectively. Above 100 GHz free space methods must be used (see, for example, Chamberlain, Haigh, and Hine 1971), but these are relatively unimportant for biological solutions and will not be considered in this book.

Alternative methods can be classified as time domain techniques, generally known as time domain reflectometry (TDR) or time domain spectroscopy (TDS). With such methods a short rise-time pulse is passed into the test liquid and the input and output pulses are compared (as functions of time). By means of a suitable mathematical transformation the frequency response of the sample can be calculated. Thus in theory a complete dielectric dispersion curve can be obtained from one single measurement, although in practice more than one set of apparatus is required. Nevertheless TDS may still appear to be a very attractive technique although it has only been developed within recent years and its accuracy is at present less than that of the more traditional methods. A full description of TDS, with a discussion of its uses and limitations is given in § 3.6.

Finally sample preparation should be considered. One great advantage of the dielectric measuring technique is that very little special preparation is required compared, for example, with the difficulties encountered in preparing a specimen for electron microscopy or for an X-ray crystallographic investigation. Moreover the biological molecules can be measured in aqueous solution, i.e. their natural environment. The only special requirement is that the conductivity of the solution should be as low as possible. The techniques described here have been especially devised for liquids of high conductivity, but nevertheless in most cases the accuracy will increase as the conductivity is decreased.

3.2. Dielectric Bridges

3.2.1 *Introduction and general theory*

For the radio-frequency portion of the electromagnetic spectrum, up to say 100 MHz, complex permittivity can best be measured with an admittance bridge. The sample holder is a condenser with the test liquid forming the dielectric between its plates. Since a biological solution is conductive, the condenser is lossy and can be represented by an equivalent circuit of a resistance R in parallel with a capacitance C (Fig. 3.1).

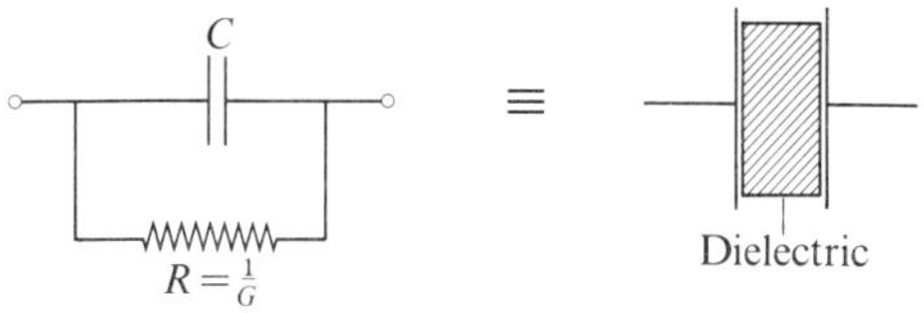

FIG. 3.1. The equivalent circuit of a lossy condenser

Although the complex impedance Z of this circuit can easily be calculated, it is more convenient to work with conductance G, defined by $G = 1/R$, and complex admittance Y, defined by $Y = 1/Z$, giving

$$Y = G + i\omega C, \qquad (3.1)$$

where $\omega = 2\pi f$, f is the frequency of the applied a.c. signal, and $i = \sqrt{-1}$. If a lossless dielectric of permittivity ε' fills a parallel plate condenser then, neglecting the distortion of the field at the edges of the plates, the capacitance will be

$$C = K\varepsilon', \qquad (3.2)$$

where K is a constant dependent upon the geometry of the system. The admittance of the condenser is given by

$$Y = i\omega C. \qquad (3.3)$$

As explained in chapter 2, $\hat{\varepsilon}$ is complex of the form $\hat{\varepsilon} = \varepsilon' - i\varepsilon''$, and if the ionic conductivity σ_s is included, then

$$\hat{\varepsilon}_T = \varepsilon' - i\left(\varepsilon'' + \frac{\sigma_s}{\varepsilon_0 \omega}\right) \qquad (3.4)$$

where σ_s is the value of the conductivity at frequencies low compared with the relaxation frequency (§ 2.3), $\hat{\varepsilon}_T$ is the total complex permittivity, and ε_0 is the permittivity of free space. Thus the lossy condenser has a complex capacitance of $\hat{C}$ where

$$\hat{C} = K\left[\varepsilon' - i\left(\varepsilon'' + \frac{\sigma_s}{\varepsilon_0 \omega}\right)\right] . \qquad (3.5)$$

Comparing eqns (3.2) and (3.3), the complex admittance will be

$$Y = iK\omega\left[\varepsilon' - i\left(\varepsilon'' + \frac{\sigma_s}{\varepsilon_0 \omega}\right)\right] . \qquad (3.6)$$

Thus, comparing eqns (3.6) and (3.1),

$$C = K\varepsilon'' , \qquad (3.7a)$$

$$G = K\left(\omega\varepsilon'' + \frac{\sigma_s}{\varepsilon_0}\right) . \qquad (3.7b)$$

At low frequencies the conductivity will often dominate eqn (3.7b) and for convenience a frequency-dependent conductivity σ can be defined as

$$\sigma = (\omega\varepsilon_0\varepsilon'' + \sigma_s); \qquad (3.8)$$

thus eqn (3.7b) becomes

$$G = \frac{K\sigma}{\varepsilon_0} . \qquad (3.7c)$$

However, care is required to ensure that confusion does not arise between σ and σ_s and in Chapter 2 we have advocated introducing a variable s defined by $s = \sigma/\varepsilon_0$. Eqn (3.7c) then becomes

$$G = Ks \qquad (3.7d)$$

in analogy with eqn (3.7a), and s has the convenient units of

THE MEASUREMENT OF PERMITTIVITY

s^{-1}. Thus if the complex admittance of the sample is measured, the permittivity ε' and the conductivity σ can be calculated from eqns (3.7a) and (3.7c), K being obtained by calibration.

Such an admittance could be measured with a Wheatstone bridge circuit (Fig. 3.2). Each of the four arms contains an impedance Z and at balance

$$\frac{Z_1}{Z_2} = \frac{Z_3}{Z_4} . \qquad (3.9)$$

For a review of various bridge networks readers are referred to the sections by Vaughan and Hill, Vaughan, Price, and Davies (1969). Practical measurements are usually made with a commercial admittance bridge. These instruments are generally based on the Wheatstone Bridge principle where Z_2, Z_3, and Z_4 are standards of impedance built into the bridge and at balance the value of the unknown admittance $Y_1 = 1/Z_1$ can be determined from calibrated dials. A particularly useful bridge is the transformer ratio arm bridge due to Calvert which is a development of earlier bridges such as those of Starr (1932) and Cole and Gross (1949).

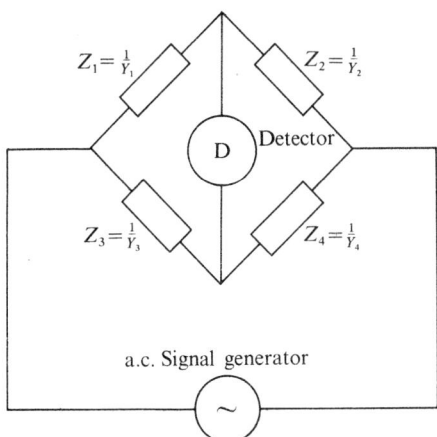

FIG. 3.2. The Wheatstone bridge circuit

Various types of sample holder may be chosen. These include the cylindrical or coaxial cell (Lovell and Cole 1959; O'Konski and Edwards 1968; Dannhauser 1970; Payne and Theodorou 1972) or a cell with two posts of circular cross-section immersed in the liquid (Pauly and Schwan 1966; Lamote, Denoo, Rosseneu-Motreff, and Peeters 1971). However, we have found the parallel plate type of cell to be ideal for the present application since with careful design the field distribution is linear within the test liquid. Suitable versions are described in the following sections.

Although the parallel plate cell is at first sight very simple, the theory is more complicated than suggested by Fig. 3.1. A more realistic circuit would consider the impedance of the connecting leads, say $Z_c = R_c + i\omega L_c$, whilst the cell itself has a residual capacitance C_0. In general a practical cell will have stray fringing field effects at the edges of the electrodes, although such effects can often be minimized with careful cell design. For conductive liquids there will be extraneous effects at the electrodes, due to electrode polarization which can often be represented by an admittance $Y_e = G_e + i\omega C_e$. The bridge, however, will measure the total admittance, Y_m say, from which the value of the admittance of the test liquid, Y_s, must be obtained. Fortunately, the above extraneous effects are frequency dependent and do not all occur together. For this reason bridge measurements can for convenience be divided into low-, medium- and high-frequency regions. This approach has been adopted and such problems as electrode polarization and the self inductance of connecting leads are only considered in relation to the frequency range where they have a measurable effect.

3.2.2. *Low-frequency bridge measurements: the problem of electrode polarization*

The difficulties in the measurement of conductive solutions are particularly acute at the lower frequencies, i.e. below about 0.5 MHz for a typical biological solution. The problems are mainly due to the formation of a layer of ions near to the electrodes (Oncley 1942; Schwan and Maczuk 1960; Mandel 1965; Rosen, Bignall, Wisse, and van der Drift 1969). This modifies the field distribution within the test liquid, causing an increase in the measured capacitance; the phenomenon is known as electrode polarization.

Another problem in measuring a conductive liquid at low frequencies is that a very small reactive term ωC may have to be determined in the presence of a large conductance G. To obtain good resolution it is necessary to have carefully designed standard components of resistance and capacitance. Because of these practical problems it is difficult to measure a biological solution at frequencies below a few kHz.

Various techniques have been proposed to cope with electrode polarization. However, no known method completely solves the problem and it must be stressed that it is advisable to reduce the conductivity of the measured solution as much as possible. Having done this, the effect of electrode polarization can be minimized by roughening of the electrode surfaces to increase their surface area. A larger reduction in electrode polarization can be obtained by depositing a layer of platinum black on to the electrodes (Schwan 1963; Takashima 1963). Caution is however required, since the platinum black might react with the test liquid.

To a first approximation electrode polarization can be considered as an impedance Z_e in series with the test liquid of admittance $Y_s = G_s + i\omega C_s$. This is shown in Fig. 3.3, together with an inductive effect L_c due to the connecting leads, and a parallel capacitance C_0 which will represent stray field effects for a three-terminal bridge. The form of the circuit of Fig. 3.3 might suggest that electrode polarization could be eliminated by calibration. For example, the concentration of a potassium chloride solution could be chosen so that its conductivity equals that of the unknown liquid. In this case the assumption is being made that the effects of the K^+ and Cl^- ions on the electrodes are the same as those of the ions in the test liquids. Although there is no positive evidence to support this assumption, experience shows that this method of correcting for electrode polarization gives reliable results (Takashima 1966). However, such a method is unlikely to succeed when the correction is large.

Another method is the use of a cell with variable electrode spacing. Such cells have been described by Broadhurst and Bur (1965), Rosen *et al.* (1969), and van der Touw and Mandel (1971);

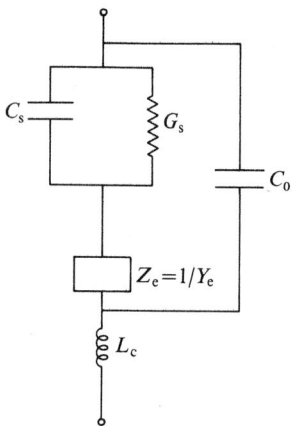

FIG. 3.3. The effect of electrode polarization in a dielectric cell

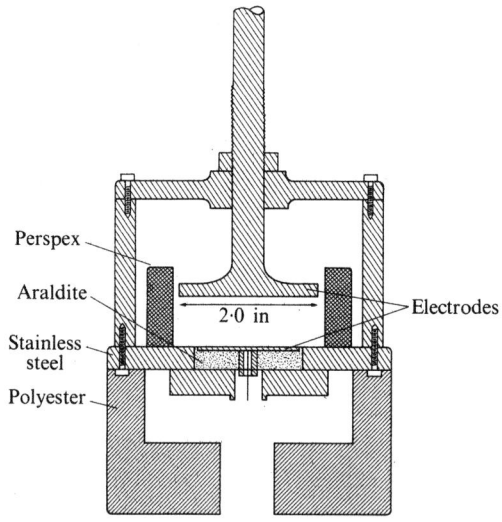

FIG. 3.4. The permittivity cell for low radio frequencies (from Young and Grant 1968)

that used by Young and Grant (1968) is shown in Fig. 3.4. This is a three-terminal electrode assembly mounted on a polyester base. A micrometer moves a stainless steel electrode of 2" diameter. In the centre of the base plate, forming the neutral electrode or shield, is a disk of 1.358" diameter. This forms the other electrode and is insulated from the shield with an epoxy resin.

The use of a movable electrode cell to correct for electrode polarization has been considered by Fricke and Curtis (1937) and more recently by Schwan and Maczuk (1960). When the electrodes are close together the polarization effect dominates, but as the separation is increased the effect of strays increases and the polarization decreases. The approach adopted by Schwan and Maczuk (1960) was to calculate the apparent permittivity of the test liquid ε'_m at various electrode separations d and then to plot log ε'_m against log d. This resulted in a graph of the type shown in Fig. 3.5 and the minimum value of ε'_m was taken as the true permittivity, ε', of the test liquid. This method has been used by Young and Grant (1968), and from tests carried out with various concentrations of potassium chloride solutions they concluded that the maximum conductivity which could be tolerated was around 3×10^{-2} Ω^{-1} m^{-1} at a frequency of a few kHz.

A similar but more complicated method requiring the use of a computer has been suggested by van der Touw and Mandel (1971)

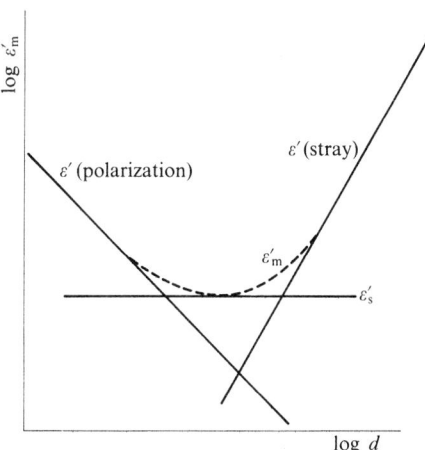

FIG. 3.5. Graph of log ε'_m against log d (for explanation of symbols see text) (from Schwan and Maczuk 1960)

and by van der Touw, Mandel, Honijk, and Verhoog (1971). They expressed electrode polarization as a power series in $1/d$ and then attempted to calculate the low-order coefficients. However, it would seem unlikely that many coefficients would be obtainable in a practical situation, so that the method would still have a very definite frequency/conductivity limit.

Other graphical methods have been proposed; see for example, Oncley (1938), Ferry and Oncley (1941) and Takashima (1963). Yet another method of dealing with the problem of electrode polarization is to use a four-terminal bridge in which two current electrodes supply the alternating field across the sample and two potential electrodes (or probes) sample the field (Ferris 1963; Berberian and Cole 1969). Provided that the potential electrodes do not take any current, the effect of polarization should be zero; however, such a system seems difficult to realize in practice, particular for very conductive solutions.

To conclude, although many methods have been proposed, there is no known ideal solution to the problem of electrode polarization. This is a case where prevention is better than cure, hence the desirability of using low-conductivity liquids wherever possible. It is vital that newcomers to the field are wary of electrode effects since even experienced workers have 'discovered' a new dispersion, only later to find that the change in capacitance with frequency was due to electrode polarization.

3.2.3 *Medium-frequency bridge measurements*

Bridge measurements in the frequency range of several hundred kHz to a few MHz should be the least difficult to obtain. At these frequencies electrode polarization effects are small or negligible, whilst with careful cell design lead inductances will not be noticeable.

A suitable cell is shown in Fig. 3.6. The liquid is contained in a perspex sample holder placed between two solid silver electrodes with roughened surfaces to minimize electrode polarization. The dimensions of the liquid holder, within the perspex, must be chosen so that the capacitance and conductance of the system are suitable for the bridge being used. It is

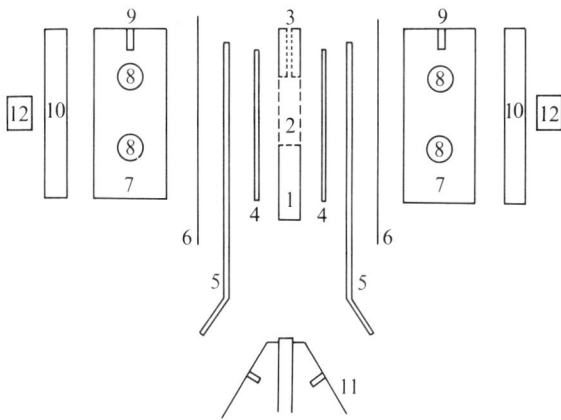

FIG. 3.6. A permittivity cell for medium radio frequencies. 1. perspex spacer, 2. cavity, 3. inlet hole, 4. silver electrode, 5. legs, 6. polythene insulators, 7. heating blocks, 8. circulating liquid, 9. thermocouple accommodation, 10. perspex heat insulation, 11. B201 terminals, 12. clamp

also important to ensure that the surface area of the electrodes is considerably greater than that of the cavity to prevent fringing effects from distorting the field within the liquid. The cell does not have to be dismantled during the calibration and measurement procedure since the liquid is introduced into the cavity, via a small hole (see Fig. 3.6), with a syringe. The electrical connections to the bridge consist of aluminium legs which sit directly on the terminals of the bridge; this avoids connecting leads and their associated resistance and self inductance. The water jacket is constructed from two metal blocks which are connected electrically to the third terminal of the bridge and insulated from the electrodes of the cell with polythene. The temperature of the system is measured with a thermocouple which can be inserted into the liquid sample. The components of the cell are all held together with a clamp; this ensures intimate contact between the various surfaces yet enables the cell to be easily dismantled for cleaning purposes.

At frequencies higher than a few hundred kHz electrode polarization will generally be negligible and the working equations are

$$G_m = G_s \tag{3.10a}$$

$$C_m = C_0 + C_s. \tag{3.10b}$$

For an 'ideal bridge' the permittivity ε' and conductivity σ could easily be obtained from the above equations and eqns (3.7a) and (3.7c). However, in practice difficulties can arise due to the limited accuracy of many commercial bridges which are normally designed for general admittance measurements not requiring high precision. In particular the calibration of the dials may be relatively inaccurate (perhaps 1–2 per cent) whilst, even more serious, there may be interaction between the C and G readings. This problem can be tackled in the following way. When a capacitance C and a conductance G are placed on to the bridge terminals, it can be assumed that

$$C = C_m + f(C_m, G_m, \omega) \tag{3.11a}$$

$$G = G_m + g(C_m, G_m, \omega), \tag{3.11b}$$

where C_m and G_m are the measured values of C and G respectively, and f and g are unknown functions caused by the extraneous factors mentioned above. The capacitances and conductances to be measured are, from eqns (3.10a), (3.10b), (3.7a), and (3.7c)

$$C = K\varepsilon' + C_0 \tag{3.12a}$$

$$G = \frac{K\sigma}{\varepsilon_0}. \tag{3.12b}$$

For a substance whose conductivity varies little with frequency (for example, a protein of relatively long relaxation time), an aqueous solution of potassium chloride can be made equal in conductivity to the test solution. Thus from eqns (3.11a) and (3.12a),

$$K\varepsilon'_s + C_0 = C_s + f_1(C_m, G_m, \omega) \tag{3.13a}$$

$$K\varepsilon'_w + C_0 = C_p + f_2(C_m, G_m, \omega), \qquad (3.13b)$$

where the subscripts s, w, and p refer to the solution, water, and potassium chloride respectively (note that $\varepsilon'_w \simeq \varepsilon'_p$ for a dilute solution). Since the conductances of the solution and the potassium chloride are approximately equal, $f_1 \simeq f_2$, leading to

$$\varepsilon'_s = \frac{1}{K}(C_s - C_p) + \varepsilon_w. \qquad (3.14)$$

By similar reasoning it can be shown that from eqns (3.11b) and (3.12b)

$$\sigma_s = \frac{\varepsilon_0}{K}(G_p - G_p) + \sigma_p. \qquad (3.15)$$

If, however, the conductivity varies with frequency then several concentrations of potassium chloride, covering the range of conductance of test solution, must be used. Alternatively calibration curves of capacitance correction against measured conductance for various frequencies can be obtained for future use.

To determine the cell constant K it is generally permissible to assume that the bridge is accurate at one frequency point (say 1 MHz) for a low-conductivity solution (this is the situation that most admittance bridges are designed to measure). Thus

$$K\varepsilon'_w + C_0 = C_w \qquad (3.16)$$

and

$$K.1 + C_0 = C_{air} \qquad (3.17)$$

since $\varepsilon'_{air} = 1$. Therefore

$$K = \frac{(C_w - C_{air})}{(\varepsilon_w - 1)}. \qquad (3.18)$$

The above work has shown that even if a commercial bridge of limited accuracy is used, with care accurate permittivity measurements can still be made.

3.2.4. *High-frequency bridge measurements: the problem of self inductance*

For bridge measurements at frequencies above a few MHz problems begin to arise owing to the self inductance of the cell and its associated leads. To a first-order approximation we can consider the effect as a self inductance L_c in series with the complex impedance of the test cell and its contents, so that the measured capacitance and conductance will be given by

$$C_m = \frac{C}{(1-\omega^2 L_c C)^2} \qquad (3.19a)$$

$$G_m = \frac{G}{(1-\omega^2 L_c C)^2} \qquad (3.19b)$$

assuming that $L_c G^2/C \ll 1$. Eqns (3.19a) and (3.19b) suggest that $\omega^2 L_c C$ should be small compared with unity, a condition which may be difficult to fulfil at high frequencies, i.e. when ω^2 becomes large. For example, to measure a capacitance of 100 pF at 100 MHz to an accuracy of ±1 per cent would require a value of L_c less than 10^{-9} to 10^{-10} Henries.

To consider the practical problems, a parallel plate cell of height h, length l and separation of plates d metres has a self inductance L of

$$L = \frac{2\pi h d}{l} \times 10^{-7} \text{ Henries.} \qquad (3.20)$$

However, the capacitance is of the form

$$C \propto \frac{hl}{d}. \qquad (3.21)$$

Thus careful cell design is required to ensure that decreasing L does not put C beyond the capabilities of the bridge, bearing in mind that a cell filled with a liquid of permittivity ε' will have a capacitance of ε' times its value when containing air. Although a decrease in h will reduce both L and C, the cell must be large enough to ensure that fringing lines of field do not enter the liquid. A useful rule of thumb is to take the length l (Fig. 3.7) to be 4 times the plate separation d. Clearly with high-frequency cells it is even more important than in the previous cases to ensure that connecting leads to the bridge are not used, thus placing additional constraints on the dimensions of the cell.

A cell, suitable for use up to at least 100 MHz with a Wayne-Kerr bridge, has been described by Jordan and Grant (1970a). A similar cell for measurements with a Boonton 33A admittance bridge up to a frequency of 50 MHz is shown in Fig. 3.7(a). The electrodes were made from stainless steel and are designed to fit directly on to the terminals of the bridge. The dielectric spacer was constructed from $\frac{1}{16}$-inch clear perspex sheet. The cavity for the test liquid could be filled with a hypodermic needle and was shaped so as to minimize the possibility of air bubbles being trapped. The cell was held together with a clamp made from perspex. The metal electrodes were hollow so that they could act as a water jacket when distilled water was passed through them. The Boonton bridge used could measure a capacitance of 120 pF at frequences up to 50 MHz but would only measure 30 pF at 100 MHz. This necessitated a special cell which is shown in Fig. 3.7(b). This cell follows the same basic design as that of Fig. 3.7(a) but it is of smaller dimensions so as to keep the total capacitance to less than 30 pF when filled with an aqueous solution.

However, even with careful cell design the assumption of negligible self inductance will give a lower accuracy than desired at the higher frequencies. This problem has been considered recently by Essex, South, Sheppard, and Grant (1975). Their approach is to consider that eqn (3.7a) should be written as

$$C = K\varepsilon' + C_0 + C_G + I, \qquad (3.22a)$$

where C is the measured capacitance, and C_0 is the residual capacitance of the system and is independent of C and G. C_G is an error term which depends upon the G reading of the bridge and I is caused by the self inductance L_c in the system. Since all the terms of eqn (3.22a) may be frequency dependent, we

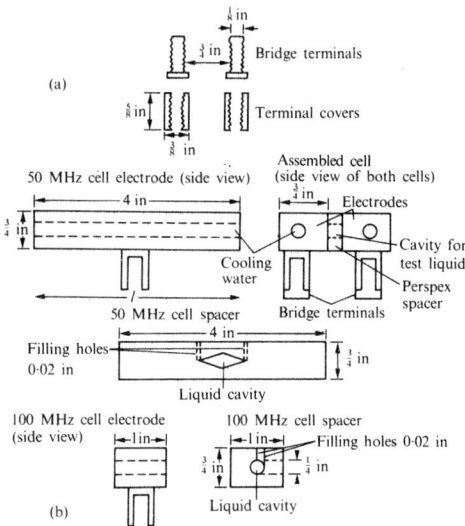

FIG. 3.7. (a) Diagram of cell suitable for measurements up to 50 MHz. (b) Diagram of cell for measurements at 100 MHz (from Essex *et al.* 1975)

have for an unknown substance of permittivity ε'_u at frequency f

$$C_u(f) = K\varepsilon'_u(f) + C_0(f) + C_G(f) + I_u(f). \quad (3.22b)$$

A solution of potassium chloride of permittivity ε'_p is then made so that its conductivity equals that of the unknown liquid at frequency f (i.e. both liquids give the same G reading on the bridge) so that

$$C_p(f) = K\varepsilon'_p(f) + C_0(f) + C_G(f) + I_p(f). \quad (3.23)$$

Thus from eqns (3.22b) and (3.23)

$$\frac{C_u(f) - C_p(f)}{K} = [\varepsilon'_u(f) - \varepsilon'_p(f)] + \frac{[I_u(f) - I_p(f)]}{K}. \quad (3.24)$$

A calibration liquid is then made by mixing various liquids which may be polar but *must not* disperse within the frequency

THE MEASUREMENT OF PERMITTIVITY

range of the bridge. The relative proportions of the liquids are determined experimentally so that the measured capacitance of the calibration liquid equals that of the unknown liquid. Potassium chloride is then added to the solution so as to match the conductance values, so that

$$\frac{C_c(f) - C_u(f)}{K} = [\varepsilon'_c(f) - \varepsilon'_u(f)] + \frac{[I_c(f) - I_u(f)]}{K} = 0. \quad (3.25)$$

It can be shown (Essex *et al.* 1975) that eqn (3.25) also implies that

$$\varepsilon'_c(f) = \varepsilon'_u(f) \quad (3.26a)$$

and

$$\frac{I_c(f)}{K} = \frac{I_u(f)}{K} \quad (3.26b)$$

From eqns (3.24) and (3.25)

$$\frac{C_c(f) - C_p(f)}{K} = [\varepsilon'_c(f) - \varepsilon'_p(f)] + \frac{[I_c(f) - I_p(f)]}{K}. \quad (3.27)$$

At the low-frequency end of the bridge (1 MHz) the self inductance is negligible, i.e. $I_c(1) - I_p(1) = 0$, so that eqn (3.27) becomes

$$\frac{C_c(1) - C_p(1)}{K} = \varepsilon'_c(1) - \varepsilon'_p(1), \quad (3.28)$$

and since the calibration liquid does not disperse between 1 MHz and any higher frequency f,

$$\varepsilon'_c(f) = \varepsilon'_c(1) \quad (3.29)$$

and

$$\varepsilon'_p(1) = \varepsilon'_p(f) = \varepsilon'_w(f). \quad (3.30)$$

Thus from eqns (3.29) and (3.30), eqn (3.28) becomes

$$\frac{C_c(1) - C_p(1)}{K} = \varepsilon'_c(f) - \varepsilon'_w(1), \qquad (3.31a)$$

which from eqn (3.26a) becomes

$$\frac{C_c(1) - C_p(1)}{K} = \varepsilon'_u(f) - \varepsilon'_w(1). \qquad (3.31b)$$

This shows that the difference in permittivity between an unknown liquid and water at a frequency f can be determined from measurements at 1 MHz only.

The cell constant is determined from

$$K = \frac{C_w - C_a}{\varepsilon_w - 1}, \qquad (3.32)$$

where C_a is the capacitance of the cell when filled with air. The conductivity σ can be determined in an analogous fashion from

$$\frac{G_c(1) - G_p(1)}{K} = \frac{\sigma_u(f) - \sigma_w(1)}{\varepsilon_0}. \qquad (3.33)$$

The above procedure for capacitance is illustrated diagrammatically in Fig. 3.8. Although this should be self explanatory, more details can be found in Essex *et al.* (1975).

For the above method to succeed a bridge with good discrimination is required, but absolute accuracy is unimportant. Although the calibration liquid has to match the unknown in both C and G at each high frequency, the method is simple and can, when used with care, give high precision results. Other methods have been proposed (Lamote *et al.* 1971; Lumry and Yue 1965) but these tend to be difficult to apply, requiring either graphical methods or the evaluation of long algebraic expressions.

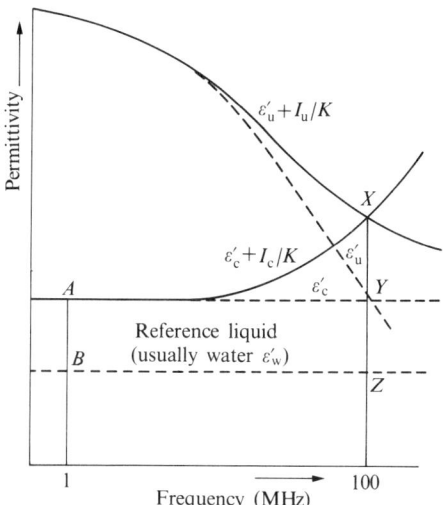

FIG. 3.8. Calibration procedure to correct for self inductance (from Essex et al. 1975)

3.3. Coaxial lines

3.3.1. *Introduction and basic theory*

Beyond a few hundred MHz dielectric measurements can no longer be made with a.c. bridges and alternative techniques such as transmission lines are employed. For the present application coaxial line systems are the most suitable for the frequency range of 50 MHz to around 12 GHz.

For readers who are not well acquainted with coaxial line techniques a few introductory paragraphs are given. The experimental cell consists of a coaxial line, i.e. a hollow metal cylinder with a concentric inner conductor. At the end of such a line is a flat metal plate which acts as a short circuit (see Fig. 3.9). An electromagnetic wave travels through the liquid and is reflected by the short circuit. This reflected wave will interfere with the incident wave so that a standing wave is set up within the liquid. It then follows that the permittivity of the liquid depends upon the shape and size of the standing wave and can be calculated from it.

Although the theory of such a system could be derived in terms of the impedance per unit length of the liquid dielectric, it is found that at the high frequencies usually associated

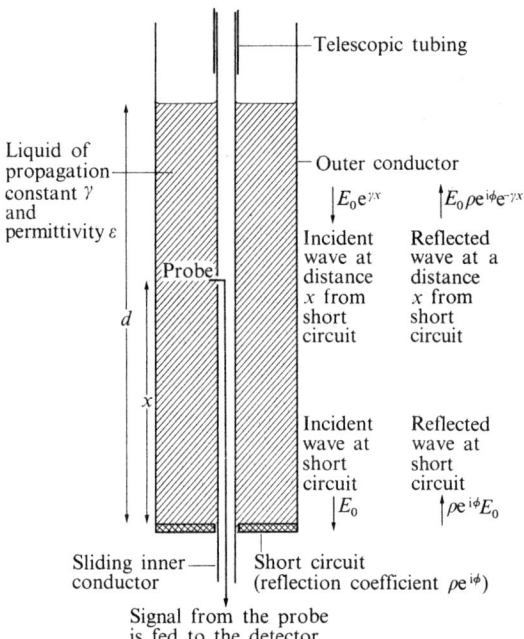

FIG. 3.9. A simplified coaxial line cell

with coaxial lines (about 100 MHz), the concepts of current and potential have less physical meaning than at the lower bridge frequencies. Thus the theory of coaxial systems tends to follow a wave approach in which the conductors of the coaxial line act as boundaries for an electromagnetic field propagating through the dielectric medium. The theory of such a system involves the solution of Maxwell's equations, subject to the appropriate boundary conditions (Lamont 1959; Stratton 1941) and leads to the concept of a complex propagation constant γ. If for a coaxial line, at a point $x = 0$, the electric component of the field is ε_0, then at a point $x = d$ the field E in the absence of reflections will be given by

$$E = E_0 e^{-\gamma d}, \qquad (3.34)$$

where $\gamma = \alpha + i\beta$, α is the attenuation coefficient, i.e. the attenuation per unit length, and β the phase constant. The units of α are either neper cm^{-1} defined by

$$\alpha_{(\text{nep cm}^{-1})} = \frac{1}{x} \log_e\left(\frac{V_1}{V_r}\right), \qquad (3.35)$$

where V_1 is the voltage at a distance x from a reference voltage V_r, or decibel cm^{-1} (db cm^{-1}) defined by

$$\alpha_{(\text{db cm}^{-1})} = \frac{1}{x} 20 \log_{10}\left(\frac{V_1}{V_r}\right). \qquad (3.36)$$

From eqns (3.35) and (3.36)

$$\alpha_{(\text{nep cm}^{-1})} = 0.1151 \, \alpha_{(\text{db cm}^{-1})}. \qquad (3.37)$$

Confusion can however arise, since under the SI system of units dimensionless quantities, such as the neper or the decibel, should not be quoted, and thus both α neper cm^{-1} and α db cm^{-1} have units of cm^{-1}. In order to avoid such problems in this chapter, the units of α will be given as neper cm^{-1} or db cm^{-1}.

At coaxial line frequencies another useful parameter is the wavelength of the radiation λ. Conventionally the wavelength in free space (or air) is denoted by λ_a and that in a dielectric medium by λ_m. The phase constant is related to the wavelength in the medium by

$$\beta = \frac{2\pi}{\lambda_m}. \qquad (3.38)$$

Essentially the coaxial line techniques to be described will measure the propagation constant γ, i.e. (α, β) or (α, λ_m). This then enables the complex permittivity ε to be determined from

$$\gamma = \frac{2\pi}{\lambda_a}\left[\left(\frac{\lambda_a}{\lambda_c}\right)^2 - \hat{\varepsilon}\right]^{\frac{1}{2}}, \qquad (3.39a)$$

where λ_c is known as the cut-off wavelength. For the principal mode in a coaxial line, λ_c is infinite and thus eqn (3.39a) simplifies to

$$\gamma = \frac{2\pi}{\lambda_a}\left[-\hat{\varepsilon}\right]^{\frac{1}{2}}. \qquad (3.39b)$$

The term principal mode will be explained in § 3.3.2; however, at this stage it can be thought of as the main form of propagation of an electromagnetic wave in a coaxial line.

The above equation is complex and when expanded into real and imaginary parts gives

$$\varepsilon' = \frac{\lambda_a^2}{4\pi^2}(\beta^2 - \alpha^2) \qquad (3.40)$$

$$\varepsilon'' = \frac{\alpha\beta\lambda_a^2}{2\pi^2}, \qquad (3.41)$$

thus showing that ε' and ε'' are functions of both α and β, although ε' depends strongly upon β and ε'' upon α. For a lossless liquid ($\alpha = 0$) eqns (3.38) and (3.40) give

$$\lambda_m = \frac{\lambda_a}{\sqrt{\varepsilon'}}, \qquad (3.42)$$

which is the equation well known from optical theory.

One design of experimental cell consists of a coaxial line terminated with a short circuit upon which the test liquid is placed. The reflection of the wave by the short circuit causes a standing wave to be set up. Some techniques then observe the reflection from the liquid sample in an air-filled line; see for example, Roberts and von Hippel (1946), Schwan and Li (1953) and Payne and Theodorou (1971). However, for the present application to high-loss liquids we find it better to observe the standing wave *within* the liquid. This is achieved by having a probe, protruding from the inner conductor, which is able to move through the liquid. A suitable cell is described in detail in § 3.3.3 and the basic principles are shown diagrammatically in Fig. 3.9. The equation of the electric field component at the probe is

$$E = E_0(e^{\gamma x} - \rho e^{i\varphi} e^{-\gamma x}), \quad (3.43)$$

where E_0 is the incident field at the short circuit of complex reflection coefficient $\rho e^{i\varphi}$. Such a standing wave is shown in Fig. 3.10; note however that this is shown in terms of decibels, rather than the amplitude of the electric field (see eqn 3.47).

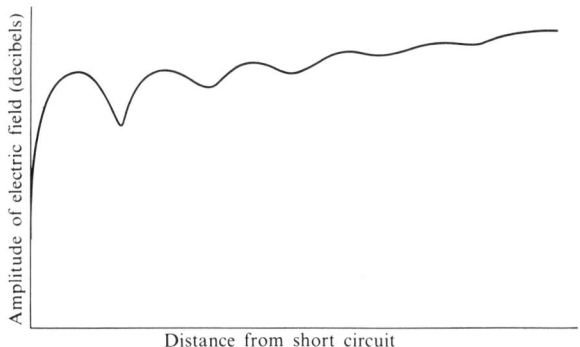

FIG. 3.10. Variation of electric field with distance from the short circuit in a coaxial line containing a lossy liquid

For an aqueous solution the dipolar components of the attenuation coefficient α will increase with increasing frequency throughout the coaxial line range, and beyond 2 GHz the standing wave becomes ill defined. Hence at frequencies higher than this a change from standing to travelling wave technique is required. It can be assumed that for a large α a purely travelling wave exists in the top portion of the cell, i.e. the equation of the field is $E = E_0 e^{-\gamma x}$. The cell can be put into one arm of a microwave bridge and either by successive balancing of the bridge or by observing the output signal of the bridge as it goes off balance, γ can be calculated. Suitable standing and travelling wave techniques are fully described in the following sections.

3.3.2. *Higher modes*

The theory of a coaxial line system is more complicated than suggested by the previous section, where only the principal or transverse electromagnetic mode (TEM) was considered. This mode is unique since it can propagate freely at all frequencies and has both its electric and magnetic field components perpendicular to the direction of propagation. However, the solution of Maxwell's equations shows that two infinite series of modes can also exist, namely the H or transverse electric (TE) and the E or transverse magnetic (TM). With H modes the electric field component, and with E modes the magnetic field component, is perpendicular to the direction of propagation. The various modes within the two series are denoted by H_{mn} and E_{mn} respectively. For further details concerning Maxwell's equations and their solution, readers are referred to standard texts such as Stratton (1941), Lamont (1959), and Baden Fuller (1969).

If modes other than the TEM are present, then the results obtained from the cell cannot be interpreted. Fortunately for an air-filled line each higher mode will only propagate freely for wavelengths λ less than a critical wavelength λ_c and will decay exponentially for $\lambda > \lambda_c$. Although the exact calculation of λ_c requires the solution of equations containing Bessel Functions, the following approximations (Marcuvitz 1951) are generally sufficient. For an E_{mn} mode

$$\lambda_c \simeq 2(b-a)/n \qquad n = 1, 2, 3; \qquad (3.44)$$

for an H_{m1}

$$\lambda_c \simeq \pi(a+b)/m \qquad m = 1, 2, 3; \qquad (3.45)$$

and for an H_{mn} mode

$$\lambda_c \simeq 2(b-a)/(n-1) \qquad n = 2, 3, 4. \qquad (3.46)$$

a and b are the radii of the inner and outer conductors respectively. The substitution of λ_c into eqn (3.39a) gives γ for any required mode and shows that even if higher modes are generated by the microwave oscillator, the dimensions of coaxial line components can be chosen so that no higher modes reach the experiment. However, discontinuities within the experimental system will give mode conversion, i.e. some of the TEM mode is converted into higher modes. Although the effect of discontinuities can be calculated numerically (Wexler 1969), a computer programme is required. Nevertheless mode conversion can be visualized pictorially (see Fig. 3.11). The boundary conditions for a coaxial line require that the electric lines of field are perpendicular to the metallic conductors and thus changes in the geometry of the system must distort the field. This distortion cannot be represented by the TEM mode and thus mode conversion occurs.

This shows that an important requirement of good cell design is that of regular geometry coupled with the minimum of discontinuities.

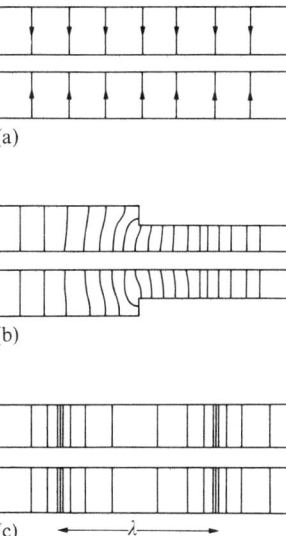

FIG. 3.11. The formation of higher modes: (a) d.c. applied to a coaxial line, (b) d.c. applied to a coaxial line with discontinuity, (c) a.c. applied to a coaxial line without a discontinuity

3.3.3. *Design of a coaxial line cell*

The upper frequency limit of a coaxial line cell will depend mainly upon the radii of its conductors. Considering the higher mode with the lowest attenuation coefficient (the H_{11}) then, knowing the dimensions of the cell, its cut-off wavelength λ_c can be estimated from eqn (3.45). For a cell containing a lossless liquid ($\varepsilon'' = 0$) of permittivity ε', eqn (3.39a) shows that for $[(\lambda_a/\lambda_c)^2 - \varepsilon']^{\frac{1}{2}}$ real, i.e. $\lambda_a > \lambda_c/\sqrt{\varepsilon'}$, then only the TEM mode can be propagated. Thus the cell can be operated up to a frequency of $c\sqrt{\varepsilon'}/\lambda_c$ where c is the velocity of light.

However, for a cell containing a high-loss liquid the upper frequency limit is not well defined since eqn (3.39a) is now complex. Thus there is no sharp transition in the propagation constant of a mode near to a wavelength of $\lambda_c/\sqrt{\varepsilon'}$. This means that the attenuation coefficient of the H_{11} mode may not be much greater than that of the TEM mode. Although in theory any required mode separation can be achieved if the cell is sufficiently small, in practice mechanical considerations will put a lower limit on the size of the conductors. Thus since mode filtering cannot be achieved solely by a suitable choice of dimensions, great care is required to ensure that the cell has symmetrical conductors and a minimum of discontinuities to reduce the likelihood of higher modes actually being generated (see § 3.3.2).

A cell suitable for measurements on aqueous solutions up to 4 GHz is shown in Fig. 3.12; it evolved from a design of Buchanan (Buchanan and Grant 1955) and was further modified by Grant and Keefe (1968). The radii of outer and inner conductors are 0.355 and 0.197 cm respectively; for a detailed account of the choice of cell dimensions, see Sheppard and Grant (1972), and Sheppard (1971). The essential features of the cell are shown in Fig. 3.12(a). The middle section contains the test liquid and is composed of an outer silver conductor and a hollow Monel metal inner conductor. A small platinum probe protrudes by about 0.1 mm from a hole in the inner conductor from which it is electrically insulated with Araldite. A piece of insulated single core wire is soldered to the probe and passes through the centre of the inner conductor.

THE MEASUREMENT OF PERMITTIVITY

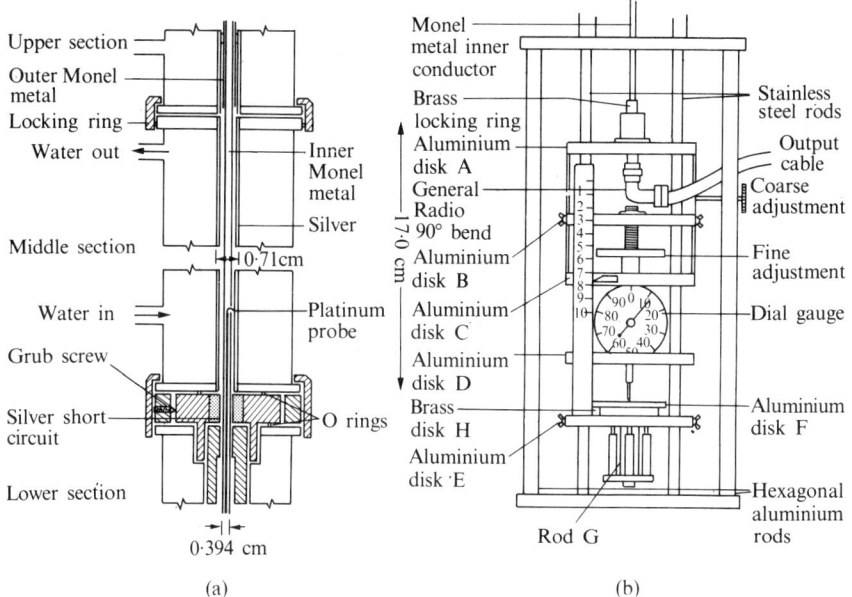

FIG. 3.12. A coaxial line cell: (a) cross section of the cell, (b) the base unit (from Sheppard and Grant 1972)

The upper section of the cell has Monel metal telescopic tubing as its inner conductor so that the inner conductor of the middle section can slide freely within it and yet maintain constant electrical continuity. The lower end of the outer Monel metal is tapered to minimize the discontinuity at this critical point.

The lower section of the cell contains a silver short circuit set into a brass housing. A hole is drilled in the centre of the silver and then lapped down so that the inner conductor can just slide through it. These three basic sections of the cell are surrounded with a water jacket for controlling temperature.

The cell sits on a solid base unit containing the mechanical movement and measuring apparatus (Fig. 3.12(b)). This will not be described here since it was designed for a specific application and many other alternatives are possible; it is fully documented by Sheppard and Grant (1972). For the automated systems to be described, the fine adjustment facilities and even the dial gauge (Fig. 3.12(b)) would not be required and could be replaced with a mechanized drive unit.

Above 4 GHz it is better to disperse with the probe type of cell. Payne and Theodorou (1971) have designed a cell for use up to 9 GHz which relies on the observation of a standing wave external to the liquid. Although their cell can be varied in length, this is only done to produce optimum working conditions, the length being held constant during the measurement procedure. At present we are designing a coaxial line cell for use up to 12 GHz which will enable the length of the test liquid column to be varied as part of the experimental process; the details of this cell will be published in due course.

For work on aqueous solutions at lower frequencies a cell has been described by Lawinski, Shepherd, and Grant (1975). This works on a similar principle to that described above and enables measurements to be made at frequencies down to 50 MHz.

3.3.4. *Methods of measurement not requiring a computer*

Although a computer analysis of coaxial line data is recommended a brief résumé of older methods of analysis is given.

(a). Standing wave. A standing wave circuit is given in Fig. 3.13 and shows a superheterodyne detection system. The signal generator of frequency f_0 MHz feeds the signal through a low-pass filter, an isolator, and then into the coaxial line cell. The filter attenuates harmonics of the fundamental frequency whilst the isolator prevents any of the signal from being reflected back into the oscillator. An isolator can be

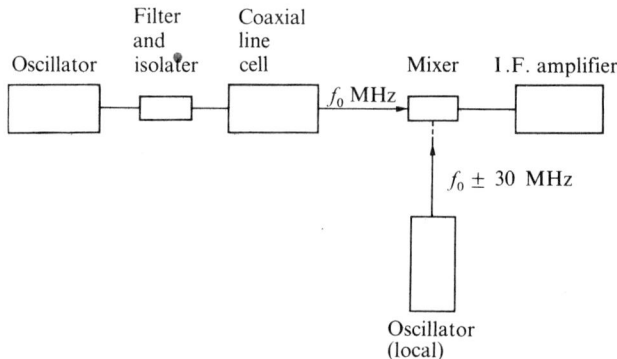

FIG. 3.13. A basic circuit for standing wave measurements

inconvenient, owing to its limited frequency range of operation, and is often replaced with an attenuator of 6 to 10 db. The output from the cell is fed into a mixer where it combines with a signal from the local oscillator of frequency $f_0 \pm 30$ MHz. The resulting 30 MHz intermediate frequency (IF) is then amplified and the output displayed on a meter calibrated in decibels.

The equation of the standing wave (3.43) expressed in decibels is

$$db_{obs} = db_0 + 20 \log_{10} |(e^{\gamma x} - \rho e^{i\varphi} e^{-\gamma x})|, \qquad (3.47)$$

where the negative sign denotes a phase change of π radians on reflection. Under the assumption that the short circuit is perfect (i.e. $\rho = 1$, $\varphi = 0$), then for a very low-loss liquid ($\alpha^2/\beta^2 \ll 1$) the distance between two successive minima of the standing wave equals $\lambda_m/2$. If the amplitude ratio between a maximum and a minimum is r then

$$r = \sinh(\alpha x_1)/\cosh(\alpha x_2), \qquad (3.48)$$

where x_1 and x_2 are the distances from the short circuit of a minimum and maximum respectively. Thus by a method of successive approximations α can be calculated from eqn (3.48). As the ratio α^2/β^2 increases, the approximations possible to eqn (3.47) become less and eqn (3.48) increases in complexity. Although the computation of α and β is still possible without a computer, the method can be tedious and has only been used extensively in the pre-computer era (Buchanan and Grant 1955).

For applications of limited computer access, tables can be generated which enable α and β to be directly determined (Shepherd 1967; Barbenza 1969).

(b) Travelling wave. As mentioned in the introduction, the standing wave method cannot be used for very high-loss liquids and the technique must change to that of travelling wave. The circuit is given in Fig. 3.14. The source and detection system are the same as for the standing wave (Fig. 3.13) but the circuit now contains a microwave bridge. Measurements are only taken from the top portion of the liquid where any signal

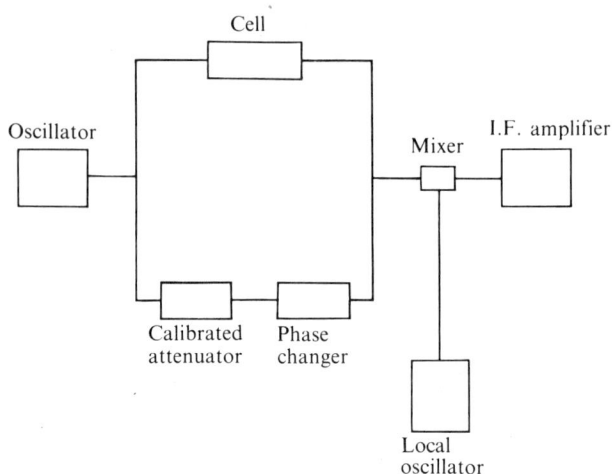

FIG. 3.14. A basic circuit for the travelling wave method

reflected from the short circuit is negligible. It should be possible to move the probe of the cell through the liquid and to successively adjust the calibrated phase shifter and attenuator so as to balance the bridge, thereby obtaining α and β (or λ_m). However, the following method gives greater accuracy. The phase shifter and attenuator are disconnected, i.e. the circuit is essentially that of the standing wave. A plot of decibels against the distance of the probe from the top of the cell gives a straight line of slope α, this being known as an α-plot. The bridge is then reconnected and successively balanced by moving only the probe and the attenuator, i.e. the phase is not changed. The distance in the cell between two balance positions is then equal to λ_m. This method has been used successfully by Buchanan and Grant (1955) and by Pennock and Schwan (1969).

The limit of the methods occurs when the attenuation of the liquid is so large that the second balance cannot be obtained. Under these conditions it is possible to use a phase shifter quantitatively (Grant and Sheppard 1970).

3.3.5. *Non-automated computerized techniques*

(a) *Standing wave.* When a computer is available measurements of the standing wave can be taken at regular intervals along the cell rather than at the maxima and minima. These data can then be fitted to eqn (3.47) with a least-squares minimization program (Jordan and Grant 1970b) (see Chapter 4 for details). However, it is better to use eqns (3.47) and (3.39b) to give

$$db_{obs} = db_0 + 20 \log_{10}\left\{|\exp[(2\pi/\lambda_a)\sqrt{\hat{\varepsilon}}(x-x_0)] - \rho\exp[(-2\pi/\lambda_a)\sqrt{\hat{\varepsilon}}(x-x_0)]|\right\}. \quad (3.49)$$

This is a five-parameter fit where $\hat{\varepsilon} = \varepsilon' - i\varepsilon''$ is the complex permittivity, db_0 is a constant, ρ the modulus of the complex reflection coefficient, and x_0 the position of the short circuit.

Apart from the avoidance of a tedious calculation this method has many advantages. In particular more data points can be taken than with methods restricted to maxima and minima. Measurements can be taken at lower frequencies, since even if the first minima cannot be observed reasonable parameter estimates can be obtained. The errors of, and the correlation coefficients between, the parameters can easily be estimated (see Chapter 4).

Although the computer must be provided with initial estimates of the parameters this is easily accomplished. Often ε' and ε'' will be known approximately, and if not α and λ_m can be estimated from the standing wave. The constant db_0 can be taken as a guessed average of the maxima and minima whilst ρ and x_0 can be set at 1 and 0 provided that the short circuit position is known.

(b) *Travelling wave*. A computer is not necessary for travelling wave measurements. However, if available it can be useful for calculating α from α-plot data, particularly if an accurate estimation of errors is required. It is also helpful in calculating ε' and ε'' from α and λ_m, particularly when wall loss and conductivity corrections are required (see § 3.3.7).

3.3.6. *Automated computerized techniques*

Since the dielectric investigation of a substance requires much experimental work, some form of automatic data acquisition is desirable. This can be achieved by means of either a data logging system or a small laboratory computer, the data being punched automatically on to computer tape.

(a) Standing wave. The circuit for the automated standing wave is shown in Fig. 3.15, where it can be seen that only one microwave oscillator is required (see Sheppard (1972) for reasons why the superheterodyne method was rejected). This microwave oscillator is modulated with a 1 kHz square wave; the resultant modulated output from the detector is passed to an amplifier and then to a phase-sensitive detector (PSD). A PSD is essentially a tuned amplifier which locks on to the modulated frequency, thus improving the signal-to-noise ratio of the system without the problems of frequency drifting usually associated with tuned amplifiers. Another departure from the previous methods is that the microwave signal is divided with a tee, before the cell, so that the output of the oscillator can be monitored. Although this division could be accomplished with a directional coupler, this is a narrow-band item and thus not suited to the broad-band coaxial line. However, since a tee gives an equal division of power, a 40-db attenuator is used to protect the diode of the monitor channel.

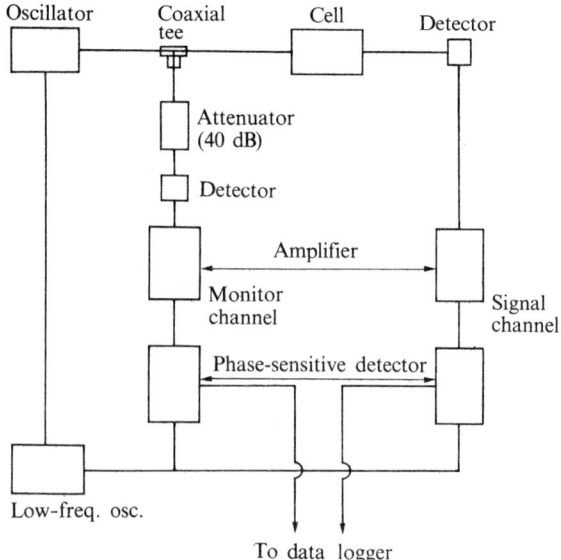

FIG. 3.15. A circuit for the automated standing wave technique (from Sheppard 1972)

The output signals of the PSDs are fed to a digital voltmeter which forms part of either a data logging or an on-line computer system. The probe of the experimental cell is driven with a stepping motor. At each position the data logger or computer measures and records the output of the monitor and signal channels, and then commands the stepping motor to increment.

Although the data are analysed essentially as for the non-automated system, two additional points must be considered. With this system the output signal is measured in volts, unlike previous methods where decibels have been used. Whether voltage or decibels are used in the minimization program is not a matter of arbitrary choice. It has been shown (Sheppard 1972) that in order to satisfy the requirement that the expected variances of the experimental results should be equal (see Chapter 4), decibels should be minimized. Thus the computer program converts the voltage readings to decibels before minimization. Eqn (3.49) will not be valid if the diode is not a perfect square-law detector. Under the assumption that a simple power-law relationship holds over the range of measurements, eqn (3.49) can be written as

$$db_{obs} = db_0 + 10n\log_{10}\left\{|\exp[(2\pi/\lambda_a)\sqrt{\varepsilon}(x-x_0)] - \rho\exp[(-2\pi/\lambda_a)\sqrt{\varepsilon}(x-x_0)]|\right\}, \quad (3.50)$$

where n is the characteristic of the crystal and will be close to two. Thus a six-parameter minimization is required.

(b) Travelling wave. The circuit for the automated travelling wave is shown in Fig. 3.16 and is essentially the bridge circuit of Fig. 3.14 with the detection system of Fig. 3.15. It was decided that, rather than attempting to design a bridge which would balance automatically, it would be better to observe the output of a standard microwave bridge as it goes off balance.

For a 'perfect' bridge the output signal in decibels is

$$db_{obs} = db_{ref} + 10n\log_{10}\left\{|E_s e^{i\theta_s} e^{-\gamma x}+1|\right\}, \quad (3.51)$$

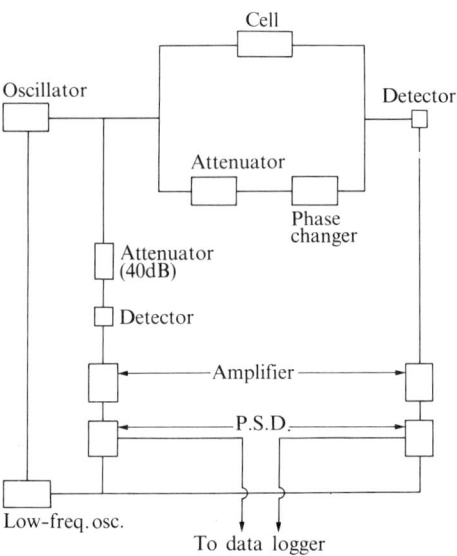

FIG. 3.16. The circuit for the automated travelling wave method (from Sheppard 1972)

where db_{ref} is the signal through the arm of the bridge containing the calibrated phase shifter and attenuator; E_s and θ_s are proportional to the amplitude and phase of the signal through the cell, at $x = 0$, whilst n is the characteristic of the crystal. However, in a practical situation interaction will occur between the two arms of the bridge and can be represented to a first-order approximation (Sheppard 1972) by

$$db_{obs} = db_{ref} + 10n\log_{10}\left\{\left|\frac{E_s e^{i\theta_s} e^{-\gamma x} + 1}{E_I e^{i\theta_I} + 1}\right|\right\}, \qquad (3.52)$$

where $E_I e^{i\theta_I}$ accounts for the interaction.

In practice the probe is put to the bottom of the cell and since for a high-loss liquid $e^{-\gamma x} \to 0$, db_{ref} is directly measured. For convenience the bridge is initially balanced (this need not be an accurate balance) so that the estimates of E_s and θ_s are 1 and π respectively. Data can now be recorded automatically, as for the standing wave, and then fitted to eqn (3.52). The initial point at which the bridge was balanced is, however, not recorded since this must have a bad signal-to-noise ratio.

Although all of the required parameters can be obtained by the above method, α can be calculated more accurately from a separate α-plot (see § 3.3.4b). The slope of the α-plot will depend upon the characteristic of the crystal n. This can, however, be obtained from the computer analysis of eqn (3.52) so as to yield a corrected value for α.

3.3.7. *Corrections for wall loss and ionic conductivity*

It is important that the values of permittivity obtained by all of the previous methods be corrected for the ionic conductivity of the solution and the wall losses of the experimental cell. Although SI units are mainly used in this book, it is traditional at coaxial line frequencies to express quantities in centimetres rather than metres and this convention will be followed.

The total measured permittivity $\hat{\varepsilon}_T$ is given by

$$\hat{\varepsilon}_T = \varepsilon' - i\left(\varepsilon'' + \frac{\sigma}{\varepsilon_0 \omega}\right), \qquad (3.53)$$

where ε'' is the dipolar loss and σ_1 the low-frequency ionic conductivity. The bracket can be conveniently re-written to give

$$\varepsilon_T'' = \varepsilon'' + 60\lambda_a \sigma, \qquad (3.54)$$

where λ_a is the free space wavelength in cm and σ the low-frequency conductivity in Ω^{-1} cm^{-1} (or m and Ω^{-1} m^{-1} respectively). Thus if σ is measured on a conductivity bridge ε'' can be calculated, on the assumption that σ is not frequency dependent. It has, however, been suggested that this may not be true (Falkenhagen 1934; Little and Smith 1955). If it is suspected that the conductivity at microwave frequencies may not equal its d.c. value, then σ can be fitted as a parameter in the final analysis of the data (see Chapter 4).

Many books (see, for example Jackson 1964), consider that to a first approximation wall loss can be accounted for by correcting only the attenuation coefficient α, the appropriate formula being

$$\Delta\alpha = \frac{1}{4\log_e\left(\frac{b}{a}\right)}\left(\frac{\varepsilon'}{30\lambda_a}\right)^{\frac{1}{2}}\left[\frac{1}{a\sigma_a^{\frac{1}{2}}} + \frac{1}{b\sigma_b^{\frac{1}{2}}}\right], \quad (3.55)$$

where a and b are the radii and σ_a and σ_b are the conductivities of the inner and outer conductors respectively. However, it can be shown (Stratton 1941; Sheppard and Grant 1972) that both α and β should be corrected, i.e. wall losses will influence both ε' and ε''. The total permittivity is related to the propagation constant γ by

$$\gamma = \left(\frac{2\pi}{\lambda_a}\right)\left[-\hat{\varepsilon}_T\right]^{\frac{1}{2}}. \quad (3.56)$$

When wall losses are present, eqn (3.56) will become

$$\gamma = \frac{2\pi}{\lambda_a}\left(-\hat{\varepsilon}_T\right)^{\frac{1}{2}}\left\{1 - \frac{(i-1)}{4\pi\log_e\left(\frac{a}{b}\right)}\left[\frac{1}{a\sigma_a^{\frac{1}{2}}} + \frac{1}{b\sigma_b^{\frac{1}{2}}}\right]\right\}^{\frac{1}{2}}, \quad (3.57)$$

which can be written as

$$\gamma = \frac{2\pi}{\lambda_a}\left\{-\hat{\varepsilon}_T(1-A)\right\}^{\frac{1}{2}} \quad (3.58)$$

where A is a complex wall loss correction constant given by

$$A = \frac{(i-1)}{4\pi\log_e\left(\frac{b}{a}\right)}\left(\frac{\lambda_a}{30}\right)^{\frac{1}{2}}\left[\frac{1}{a\sigma_a^{\frac{1}{2}}} + \frac{1}{b\sigma_b^{\frac{1}{2}}}\right]. \quad (3.59)$$

Although when a computer is available A can be evaluated and then substituted into eqn (3.58) it is useful to consider approximate corrections to α and β for hand computation. It has been shown (Sheppard and Grant 1972) that if a real constant A' is defined by

$$A' = \frac{A}{(i-1)\lambda_a^{\frac{1}{2}}}, \quad (3.60)$$

that is

$$A' = \frac{1}{4\pi\log_e\left(\frac{b}{a}\right)\sqrt{30}} \left[\frac{1}{a\sigma_a^{\frac{1}{2}}} + \frac{1}{b\sigma_b^{\frac{1}{2}}}\right] \quad (3.61)$$

then

$$\Delta\alpha \simeq A'/2 \; \lambda_a^{\frac{1}{2}}(\alpha+\beta) \quad (3.62)$$

$$\Delta\beta \simeq A'/2 \; \lambda_a^{\frac{1}{2}}(\beta+\alpha). \quad (3.63)$$

Wall losses have been measured experimentally for waveguides (Morgan 1949; Benson 1953; Allison, Benson, and Seaman 1957; Meredith and Preece 1963) and the general conclusion reached was that the calculated wall loss is always less than the experimental value. Thus for a coaxial line it is often desirable to evaluate A' experimentally and then to use it, in conjunction with eqn (3.60), in eqn (3.58). Although the wall losses could be measured directly in an air-filled line, they are small and thus will have large percentage errors. A better method is to measure for water at low coaxial line frequencies and then to compare it with the accurately known static value (Malmberg and Maryott 1956), thus enabling the value of A' to be deduced.

With the non-computerized techniques ε' and ε_T'' are calculated and then corrected for wall loss and conductivity. However, with the methods of curve fitting, $\hat{\varepsilon}$ in eqn (3.49) can be replaced with $\hat{\varepsilon}(1-A)$ of eqn (3.58) and thus the corrections can be computed during the minimization.

3.4. Waveguides
3.4.1. *Introduction*

For our purposes present technology puts an upper limit on coaxial line apparatus of about 12 GHz. However, beyond this frequency the electromagnetic field can propagate in waveguides. Essentially a waveguide consists of a hollow metal tube through which an electromagnetic wave can pass. Thus a waveguide can be considered theoretically as being similar to a coaxial line but without an inner conductor. For this reason the theory must follow a wave approach since it is no longer possible to consider the potential across and the currents flowing within the outer and inner conductors.

The test liquid is contained in the waveguide and as for coaxial lines the propagation of the electromagnetic waves depends upon the permittivity of the liquid. Although waveguides can be obtained with various cross-sections, for theoretical reasons the usual shape is rectangular where the ratio of the sides is 2:1.

The theory shows that, unlike coaxial lines, the TEM mode cannot exist in a waveguide. The mode with the lowest cut-off frequency becomes the dominant mode, all others being considered as higher modes. This lack of a TEM mode has great practical repercussions in that it causes waveguides to operate in a narrow frequency band.

A rectangular waveguide has a cut-off wavelength of

$$\lambda_c = \frac{1}{\left[\left(\frac{m}{2a}\right)^2 + \left(\frac{n}{2b}\right)^2\right]} , \qquad (3.64)$$

where a and b are the lengths of the sides of the guide and m and n define the order of the mode (Baden Fuller 1969; Lamont 1959). Thus any waveguide system can only operate from a wavelength less than the cut-off wavelength of its dominant mode (H_{10} for a rectangular guide) to a wavelength greater than λ_c for the first higher mode. For example, a J band waveguide could in theory only operate from 12.4 to 18.0 GHz but in practice this may be restricted even further by the narrow tuning range of the oscillators. Thus most waveguide systems operate at either one fixed frequency or have a very small tuning range.

For biological solutions the situation until a few years ago was that the high attenuation coefficient effectively ruled out the possibility of using standing wave techniques. Thus the traditional systems used at these frequencies were a development of the phase-amplitude balance method as described by Buchanan (1952). However, the recent use of computerized techniques has enabled standing wave type of systems to be employed (van Loon and Finsey 1973, 1975).

As explained above, waveguides and coaxial lines are physically similar so that much of the theory of § 3.3 is applicable to the present case. The experimental techniques still determine the propagation constant γ which enables the permittivity $\hat{\varepsilon}$ to be calculated using eqns (3.39a) and (3.64). The discussion of § 3.3.2 regarding higher modes is still applicable and although good mode separation can be achieved in an air-filled guide, this is not the case when a high-loss, high-permittivity liquid is present. Thus careful cell design and construction is required in order that higher modes are not excited.

Corrections for wall loss and conductivity (see § 3.3.7) must be considered. However, these are less important at waveguide frequencies. In particular the correction for ionic conductivity will be negligible, whilst the attenuation caused by wall losses will be small compared to the loss of an aqueous solution.

3.4.2. *Traditional measuring systems*

A basic waveguide bridge circuit is shown in Fig. 3.17 and is similar to the coaxial line circuit of Fig. 3.14. At these frequencies the most popular type of oscillator to give the required output power and stability is a klystron (Baden Fuller 1969). An experimental cell for a waveguide system is in many respects easier to construct than its coaxial line equivalent since there is no inner conductor. Measurements are generally taken by varying the thickness of the liquid samples; this can be achieved by having two interfaces in the guide, one of which can move relative to the other. Such cells have been described by Grant (1959), Mungall and Hart (1957), and by Grant and Shack (1967), and are designed for transmission measurements. Provided that a sufficient thickness of liquid is present to absorb interfacial reflections, then a purely travelling wave will exist within the cell. Under these conditions the bridge of Fig. 3.17 is successively balanced and values of β and α obtained from the readings of the calibrated phase shifter and attenuator. Alternatively the method of § 3.3.4*b* may give better results. This involves the direct measurement of the wavelength in the liquid and the determination of α by

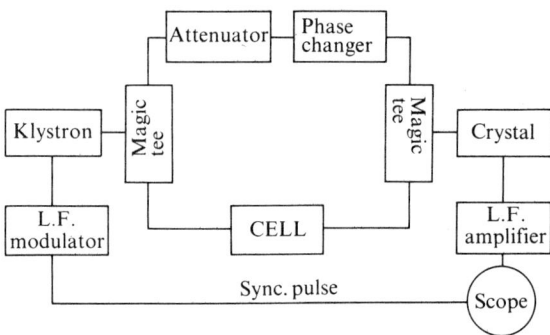

FIG. 3.17. Basic waveguide bridge circuit (from Shack 1972)

measuring the variation of amplitude with distance along the cell. It should be noted that the calculation of ε' from α and β involves the cut-off wavelength for the guide λ_c, the appropriate equation being

$$\varepsilon' = \left(\frac{\lambda_a}{\lambda_c}\right)^2 + \frac{\lambda_a^2}{4\pi^2}(\beta^2 \; \alpha^2). \qquad (3.65)$$

The equation for ε'' is unchanged (i.e., eqn (3.41)).

3.4.3. *Recent waveguide developments*

The recent introduction of computerized techniques has enabled the basic waveguide systems to be greatly improved. Clearly the apparatus of the previous section can be automated by recording the output of the bridge as it goes off balance in an analogous fashion to the coaxial line system of § 3.3.6b. When very lossy liquids are being measured it is an advantage to work with the interfaces of the cell close together, i.e. a thin liquid sample. Under these conditions a pure travelling wave will not exist in the liquid but the resultant output of the bridge can easily be analysed when a computer is present.

When aqueous solutions are being studied at the higher waveguide frequencies, problems can arise due to the very large attenuation coefficient of the water. For example at 70 GHz the wavelength in the solution is around one millimetre but the attenuation coefficient is about 35 neper cm^{-1}. Thus the ratio of α to β is so unfavourable that phase information cannot easily be extracted.

A modified bridge circuit has been described by O'Brien (1967). Essentially, after passing through the cell the signal is divided and a phase difference of 90° is introduced between the two signals. These signals are then combined with a reference signal by means of balanced mixers. Thus if the amplitude of one signal is $A \cos \theta$ then the other will be $A \sin \theta$ where A and θ are proportional to the amplitude and phase of the signal in the cell. Thus this system should be able to resolve small phase differences. Due to the necessity of working with thin liquid samples, reflections will occur between the interfaces of the cell so that a computer analysis is required. Thus the detection system of § 3.3.6b could be used, involving the modulation of the signal and its measurement with a phase-sensitive detector, coupled to a data logger or on-line computer. However, the modulation is best achieved with a ferrite modulator so that the reference signal is not modulated. Although it is felt that the rather involved circuity of the O'Brien system is best suited to frequencies above 35 GHz, it has been used by Von Casimir, Kaiser, Keilmann, Mayer and Vogel (1968) at 9.5 GHz.

Sets of waveguide apparatus have been constructed (Van Loon and Finsey 1973, 1974, 1975) to operate up to 140 GHz. A diagram of the basic system is given in Fig. 3.18(a). The microwaves travel from the klystron to the cell where they are reflected by a movable short circuit and then collected with a directional coupler. The two tuners are essentially to minimize the effect of the various reflecting interfaces within the system.

The experimental cell consists of a short circuit moving in a waveguide. An absorptive load is placed behind the short circuit so that if any power should leak it is absorbed and not reflected into what could become a resonant cavity (Fig. 3.18(b)). The question of short circuit design has also been considered by Finsey and Van Loon (1972) and Van Loon and Finsey (1974). They show that the dumbell type of short circuit

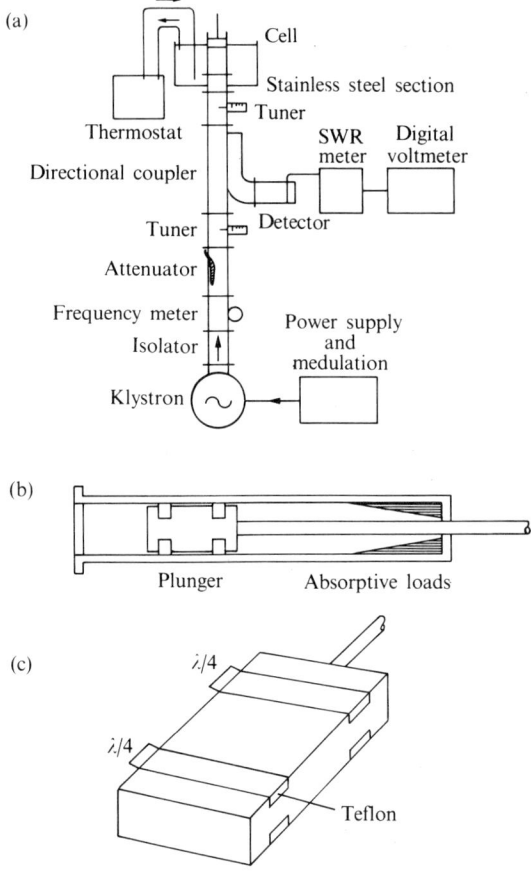

FIG. 3.18. Apparatus used by Van Loon and Finsey: (a) the experimental system, (b) the experimental cell, (c) the short circuit (from Van Loon and Finsey 1974, 1975)

advocated by Cutnell, Kranbuehl, Turner, and Vaughan (1969) could lead to higher modes, and instead advocate the design shown in Fig. 3.18(c).

As mentioned above, the major problem when measuring aqueous solutions at these frequencies concerns the retrieval of phase information. With a reflection system there are two minima per wavelength distance moved by the short circuit as compared with one for a transmission system. Clearly this is an advantage when the α/β ratio is large. However, the information contained in the experimental data is rather limited and a computer analysis is essential if this method is to succeed. For further details of the experimental arrangement and computer program, see Van Loon and Finsey (1975). With their system water has been measured up to 140 GHz, but at these frequencies the technological problems are immense due to the small dimensions of the guide and to the difficulty in obtaining sufficient microwave power. Thus it is not recommended that newcomers to the field of microwave techniques attempt to measure aqueous solutions at frequencies approaching 150 GHz.

There may be advantages in combining the simplicity of the Van Loon and Finsey system with some of the advantages of phase sensitivity possessed by the O'Brien system although we have obtained good results for water at 70 GHz by using a technique similar to that of Van Loon and Finsey (Szwarnowski and Sheppard 1977).

3.5. Other frequency domain techniques

As already mentioned it is not the aim of this chapter to give a general survey of dielectric measuring techniques but rather to concentrate upon those methods which we believe to be the best for biological measurements. However, for completeness some alternative techniques will be briefly mentioned.

Below 100 MHz a.c. bridges appear to offer the most satisfactory means of measuring permittivity accurately. However, at coaxial line and waveguide frequencies cavity resonance techniques can be used, whilst at the high waveguide frequencies there are the quasi-optical and optical techniques. The cavity methods utilize the property of a closed cavity to resonate at a frequency which depends upon the dimensions of the cavity and the permittivity ε' of the medium within it, whilst the sharpness of the resonance curve, i.e. the Q-factor, depends upon the dielectric loss ε''. Resonance techniques are ideally suited to the accurate measurement of ε'' for low-loss substances and will yield a much higher accuracy than transmission line techniques, as shown by Horikx (1970) and Stumper (1973). If, however, a high-loss liquid were to be used in the above methods then the resonance peak would be so broad that ε' could not be accurately determined. This problem can be overcome by partially filling the cavity. This is usually achieved by filling a capillary tube with the high-loss liquid and then placing it into the cavity (Collie, Hasted, and Ritson 1948; Hallenga 1972). However, such methods are more suited to observing small permittivity differences rather than to yielding absolute values.

An alternative to the above methods is the open cavity resonator whereby resonance is obtained in free space between two spherical microwave mirrors (Cullen and Yu 1971; Cullen, Nagenthiram, and Williams 1972; Bassett 1971). Although this method was essentially developed for sheets of solid dielectric it can be used for high-loss liquids contained in capillary tubes.

Beyond 150 GHz waveguides tend to be too small for the methods of § 3.4 to be successful, but as the guide dimensions increase to infinity (i.e. free space) all of the modes tend to the TEM. This principle has been utilized in overmoded techniques where an oversized waveguide is used (Garnham 1958; Taub, Hindin, Hickelmann, and Wright 1963; Carson 1970). The above methods are sometimes known as quasi-optical because of their obvious affinity with optical methods. Finally for the highest frequencies, up to about 1000 GHz, there are the optical methods where microwave interferometers are set up in free space. For details of these see Pine, Zoellner, and Rohrbaugh (1959), Vaughan, Bergmann, and Smyth (1961), Garg, Kilp, and Smyth (1965), and Davies and Chamberlain (1972). However, apart from the possibility mentioned for pure water in § 5.3.2 there is little likelihood of dispersion of the relaxation type occurring at frequencies above 100 GHz, and although resonance can occur in the high-frequency microwave or infrared regions, such phenomena are outside the scope of this book. Thus the above optical techniques are not relevent to the present application and will not be further pursued.

An alternative and new technique, although not strictly speaking in the frequency domain category, is that of thermal noise. Permittivity is an intrinsic property of a substance and thus in theory does not need an applied electromagnetic field for its investigation (see § 2.1.2). The method of thermal noise utilizes the fact that the molecules of any substance, above 0 K, are in constant motion and that this motion will in part depend upon the permittivity of that substance (Brophy and Webb 1962; Ishibashi, Sawada, and Takagi 1969). At present very little use has been made of this technique in the field of dielectrics, but the position may change in the future.

3.6. Time Domain Spectroscopy
3.6.1. *Introduction*

The previous sections of this chapter have shown how substances may have their dielectric parameters characterized by studying their permittivity at various discrete frequencies. However, it is also apparent that a large range of apparatus, plus much experimental expertise, is required. In theory it is possible to use, as an alternative, transient methods whereby a pulse is passed through or reflected from the test liquid and the input and output pulses, measured as functions of time, compared. Fourier transformation of the data then gives the frequency domain response of the system. Thus a complete frequency dispersion curve would be obtainable from the analysis of a single pulse.

Historically this method has been used for low-frequency dispersion curves (Davidson, Auty, and Cole 1951). The technique consisted in studying the growth or decay of charge across a capacitor containing a dielectric substance, the time range being from several seconds down to milliseconds. However, in order to obtain information at a frequency of a few GHz a fast rise-time pulse (of the order of a picosecond) with an equally fast detection system would be required, and it is only recently that the necessary technology has been available.

In the middle 1960s the Hewlett-Packard Company introduced a time domain system with a frequency response of up to 12.4 GHz (Oliver 1964). Initially this apparatus was used mainly by electrical engineers for making measurements on solid state devices (Davis and Loeb 1965; Nicolson 1968; Nicolson and Ross 1970). TDS was introduced into the field of liquid dielectrics by Fellner-Feldegg (1969). Although the experimental details of this paper are valuable, Fellner-Feldegg made an over-simplification in his method of data analysis. This error was noticed by Whittingham (1970) and the correct analysis was given by Fellner-Feldegg and Barnett (1970). A paper valuable for both its experimental details and methods of data analysis was published by Loeb, Young, Quickenden, and Suggett (1971). Later methods have increased the frequency range of operation, particularly at the lower end, have required less liquid, and have enabled conductive solutions to be dealt with (e.g., Fellner-Feldegg 1972; De Loor, Van Gemert and Gravesteyn 1973; Clark, Quickenden, and Suggett 1971; Suggett 1975). Various theoretical papers have also appeared which consider in more detail and extend much of the work of Fellner-Feldegg (Van Gemert 1971; Van Gemert and de Graan 1972; Van Gemert and Bordewijk 1972; Van Gemert 1974a; Giese and Tiemann 1975). There is also a useful review article by Van Gemert (1973) which, although mainly concerned with the more mathematical aspects of TDS, does contain some experimental details.

The next section describes the basic theory common to most TDS methods. The practical measuring techniques can be conveniently divided into single response methods which use relatively long cells and multiple response methods using thin cells. These methods are described in §§ 3.6.3 and 3.6.4.

3.6.2. *Basic theory*

In most TDS systems a step function rather than a pulse is applied to the test liquid. If such a step passes along a coaxial line whose dielectric changes from permittivity $\hat{\varepsilon}_1$ to $\hat{\varepsilon}_2$, then the pulse will be partially reflected and partially transmitted. When both dielectrics are non-dispersive, i.e. $\hat{\varepsilon}_1$ and $\hat{\varepsilon}_2$ are independent of frequency, then an observer at A (Fig. 3.19(a)) will see the initial pulse at time $t = 0$ and a reflection of it at time $t = t_1$, where $t_1 = 2x\sqrt{\varepsilon_1'}/c$, c being the velocity of light. Thus the arrival time of the pulse depends upon the distance of the discontinuity from A, whilst the size of the pulse gives information about the reflection coefficient ρ_{12}. Since ρ_{12} is related to $\hat{\varepsilon}_1$ and $\hat{\varepsilon}_2$ according to

THE MEASUREMENT OF PERMITTIVITY

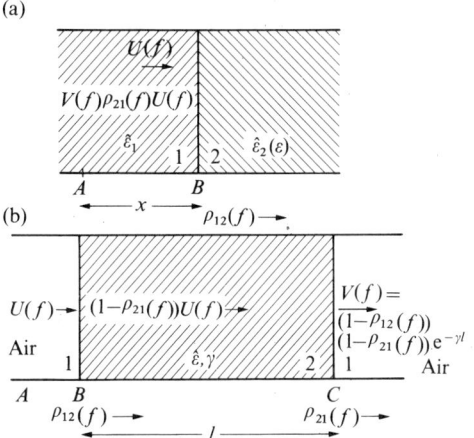

FIG. 3.19. (a) Reflection at a single interface. (b) First transmitted component through a dielectric slab.

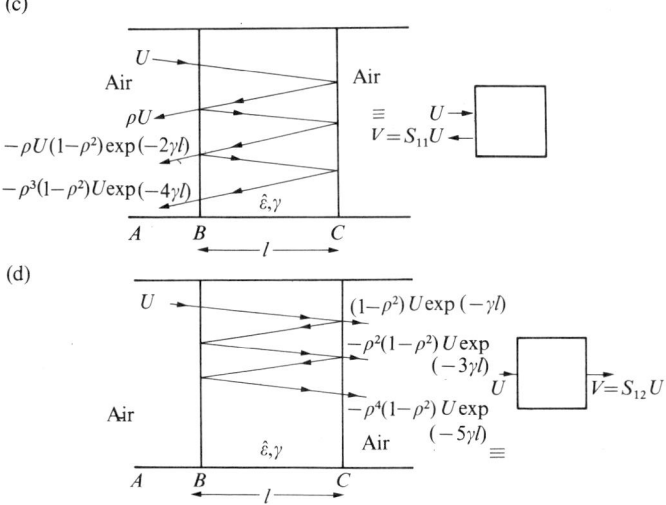

FIG. 3.19. (c) Multiple reflections and the equivalent scattering representation. (d) Total transmission and equivalent scattering representation.

$$\rho_{12} = \frac{\sqrt{\hat{\epsilon}_1} - \sqrt{\hat{\epsilon}_2}}{\sqrt{\hat{\epsilon}_1} + \sqrt{\hat{\epsilon}_2}} , \qquad (3.66)$$

where in general ρ_{12} is complex, then, if $\hat{\epsilon}_1$ is known, $\hat{\epsilon}_2$ can be calculated.

In a practical system medium 1 will be air and thus $\hat{\epsilon}_1 = 1$. However, the permittivity of the liquid may be complex and dispersive, i.e. $\hat{\epsilon}_2$ is a complex function of frequency (say $\hat{\epsilon}_2(f)$) which for simplicity of notation will be written as $\epsilon(f)$. Thus ρ_{12} is frequency dependent, so that eqn (3.66) becomes

$$\rho_{12}(f) = \frac{1 - \sqrt{\epsilon(f)}}{1 + \sqrt{\epsilon(f)}} , \qquad (3.67)$$

which results in the pulse being distorted upon reflection. This phenomenon can be understood from Fourier analysis, i.e. the pulse is considered as an infinite series of sine and cosine terms each of which is a harmonic of the basic fundamental frequency of the pulse. Since each term of the series has a different frequency, each will 'see' a different reflection coefficient at B resulting in distortion of the pulse shape. The objective of a TDS system is to observe the shape and size of the input and output pulses from which $\epsilon(f)$ can be calculated at each frequency point of interest.

To quantify the above mathematically, an input function of time $u(t)$ is being transformed to an output function $v(t)$ by the time response of the system $h(t)$. For a linear system this can be expressed by the convolution integral

$$v(t) = \int_{-\infty}^{t} u(t-t')h(t')dt' . \qquad (3.68)$$

Each of the functions u, v, and h can be expressed as functions of frequency with a Fourier transform, i.e.

$$V(f) = \int_{-\infty}^{\infty} v(t)\exp(-2\pi i f t)dt . \qquad (3.69)$$

Similarly $U(f)$ and $H(f)$ can be obtained from $u(t)$ and $h(t)$ respectively. The advantage of Fourier transformation is that eqn (3.68) becomes

$$V(f) = U(f) \cdot H(f) \qquad (3.70)$$

in the frequency domain. Thus the convolution in the time domain (eqn (3.68)) becomes a multiplication in the frequency domain (eqn (3.70)). The function $H(f)$ is known as the transfer function of the system, which from eqns (3.69) and (3.70) is given by

$$H(f) = \frac{V(f)}{U(f)} = \frac{\int_{-\infty}^{\infty} v(t) \exp(-2\pi i f t) dt}{\int_{-\infty}^{\infty} u(t) \exp(-2\pi i f t) dt} . \qquad (3.71)$$

In a practical TDS system the output and input pulses, $v(t)$ and $u(t)$ respectively, are observed and the ratio of their Fourier transforms gives the frequency response of the system, $H(f)$. For a single reflection system (Fig. 3.19(a)), $H(f) = \rho_{12}(f)$, which enables $\varepsilon(f)$ to be calculated from eqn (3.67).

For a transmission system (Fig. 3.19(b)) the transfer function $H(f)$ is

$$H(f) = [1-\rho_{12}(f)][1-\rho_{21}(f)]\{\exp(-\gamma l)\}, \qquad (3.72)$$

where from eqn (3.39b)

$$\gamma = \frac{2\pi f}{c}\sqrt{-\varepsilon(f)} . \qquad (3.73)$$

Substituting for $\rho_{12}(f)$ and $\rho_{21}(f)$ gives

$$H(f) = \frac{4\sqrt{\varepsilon(f)}}{\{1+\sqrt{\varepsilon(f)}\}^2} \exp\left\{\frac{-2\pi f l}{c}\sqrt{\varepsilon(f)}\right\}. \qquad (3.74)$$

Unfortunately eqn (3.74) cannot be solved explicitly and $\varepsilon(f)$

has to be obtained numerically using the Newton Raphson method (Hamming 1962) or some similar approach.

In a practical situation the position can be complicated by the presence of multiple reflections. However, in a TDS system each reflection will be seen as an individual event, unlike in the frequency domain where only the sum to infinity of the multiple reflections is observed. Thus it is possible to obtain only one single reflection or transmission if the cell is sufficiently long to ensure that the reflected pulses are separated in time from the initial pulse. Nevertheless it can happen that multiple reflections are inevitable or even desirable, as in the thin cell methods (see § 3.6.4).

For these situations it is useful to introduce a scattering coefficient S_{ij}. For a reflection system the *total* reflected output $V(f)$ is related to the input $U(f)$ by

$$V(f) = S_{11}(f)U(f), \qquad (3.75)$$

where $S_{11}(f)$ is the resultant of the multiple reflections within the cell BC (Fig. 3.19(c)). It can easily be seen from Fig. 3.19(c) that

$$S_{11}(f) = \rho(f)\left\{\frac{1 - \exp(-2\gamma l)}{1 - \rho^2(f)\exp(-2\gamma l)}\right\}, \qquad (3.76)$$

where $\rho_{12}(f)$ of the previous examples has been written as $\rho(f)$ and γ is given by eqn (3.73). For this situation $H(f)$ is still obtained from eqn (3.71), and then using $S_{11}(f) = H(f)$, $\rho(f)$ can be obtained by numerical methods from eqn (3.76), enabling $\varepsilon(f)$ to be calculated from eqn (3.67).

For a transmission system with multiple reflections (see Fig. 3.19(a)) the appropriate scattering coefficient is $S_{12}(f)$ given by

$$S_{12}(f) = \frac{[1-\rho^2(f)]\exp(-\gamma l)}{[1-\rho^2(f)\exp(-2\gamma l)]}. \qquad (3.77)$$

THE MEASUREMENT OF PERMITTIVITY

As before, the experiment will yield $H(f)$ which from $S_{12}(f) = H(f)$ enables $\varepsilon(f)$ to be obtained numerically.

3.6.3. *Single response methods*

These are all based on the situations shown in Figs. 3.19(a) and 3.19(b) where the samples are of sufficient length that only the first reflected or transmitted pulse is seen.

(a) Single reflection method. The basic experimental arrangement is shown in Fig. 3.20. The pulse generator, i.e. the tunnel diode, provides a step function with a rise time in the order of 35×10^{-12} s. This step then passes through the sampler to the experimental cell where it is reflected. This reflected step is then observed by the sampler which in conjunction with a sampling oscilloscope enables the pulse to be viewed. The system used by Loeb *et al.* (1971) contained two modifications. After measurements had been taken on the test liquid the cell was replaced with a short circuit. The reflected step was then taken as the input pulse, which minimizes errors due to losses within the sampler and other components. For the Fourier transformation of the data an absolute zero of time is not required, but both pulses must be matched in time so that they have the same arbitrary zero. With the Loeb system the input pulse is divided at a matched tee (see Fig. 3.20)

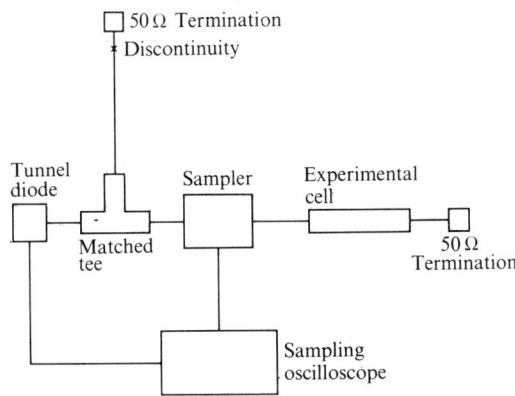

FIG. 3.20. Basic experimental arrangement for reflection measurements in the time domain (from Loeb *et al.* 1971)

to give signal and time marker pulses. The marker pulse passes down a delay line to a matched termination prior to which a slight discontinuity is introduced. This causes preferential reflection of the higher frequencies and produces a reflected spike which acts as a time reference point at the sampler.

An alternative method proposed by Loeb *et al.* is to extrapolate the linear portions of the pulses (see Fig. 3.21). The points of intersection of these extrapolations can then be matched for both the incident and reflected signals.

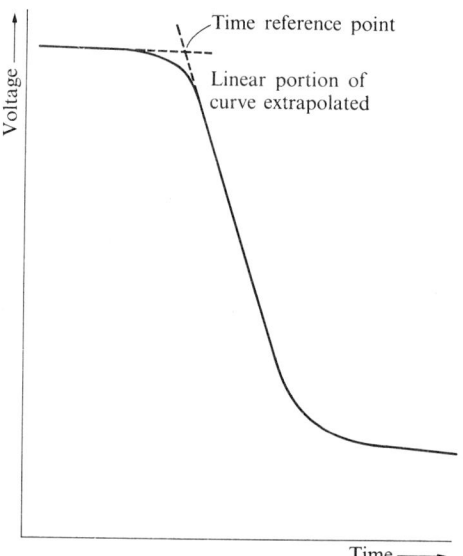

FIG. 3.21. A method for time referencing a TDS trace (from Loeb *et al.* 1971)

An idealized waveform is shown in Fig. 3.22. It can be seen that after the first reflected pulse echoes occur these limit the low-frequency response of the system. Although a long cell would separate such echoes from the first reflection, the maximum possible length for a precision cell is of the order of 30 cm. However, it is possible to obtain extra information about the dielectric from the first echo, as explained by Loeb *et al.* (1971). The basic system of Fig. 3.20 is improved if 30 cm precision air lines are placed on either side of the cell

THE MEASUREMENT OF PERMITTIVITY 109

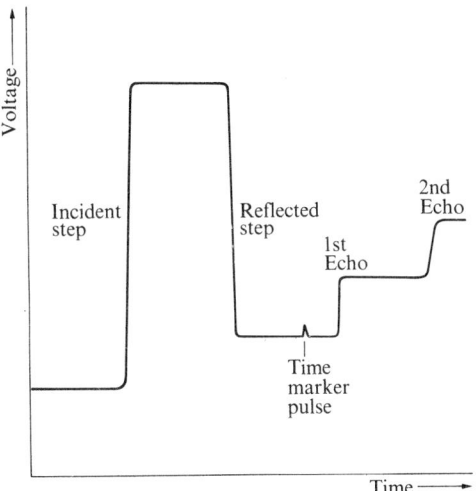

FIG. 3.22. Waveform from TDS reflection system (from Loeb et al. 1971)

so that strong reflections from possible discontinuities at the termination and the sampler are separated in time from the signal pulse.

The reflection coefficient ρ, obtained from the experiment, will be complex, leading to a complex permittivity $\hat{\varepsilon}$ (eqn (3.66)). However, errors in the phase of ρ, caused mainly by a lack of precision in time referencing, will lead to large errors in $\hat{\varepsilon}$, particularly at frequencies above a few GHz.

(b) Single transmission method. The experimental arrangement for transmission measurements is shown in Fig. 3.23. A time marker pulse is now produced by a network which preferentially passes high-frequency components. To obtain 'incident pulse' $u(t)$, the dielectric cell is replaced with an identical cell containing air. An idealized waveform from the system is shown in Fig. 3.24 and once again echoes occur. In a practical system 30 cm precision air lines would be placed at AB, CD, and EF (see Fig. 3.23).

As mentioned in § 3.6.2 it is not possible to obtain $\hat{\varepsilon}$ as an explicit function of $H(f)$ (eqn (3.74)). A method devised by Loeb et al. (1971) uses two cells of length l_a and l_b. If the ratio of the transfer functions $H_a(f)$ and $H_b(f)$ respectively is

110 THE MEASUREMENT OF PERMITTIVITY

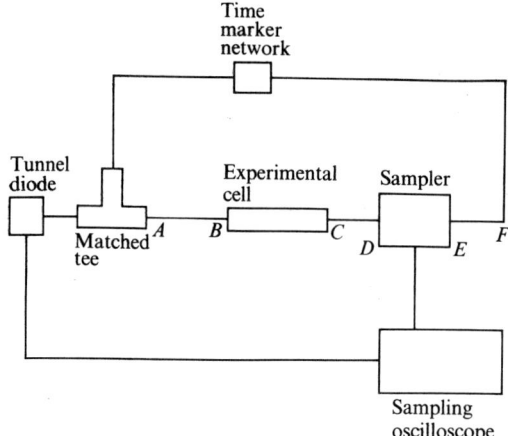

FIG. 3.23. Basic experimental arrangement for transmission measurements in the time domain (from Loeb et al. 1971)

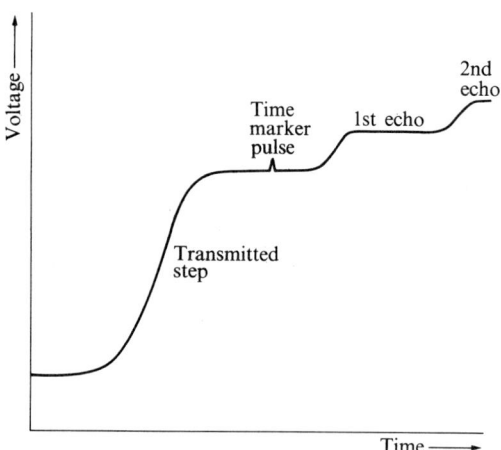

FIG. 3.24. Waveform from TDS transmission system (from Loeb et al. 1971)

denoted by

$$S \exp(i\varphi) = \frac{H_a(f)}{H_b} , \qquad (3.78)$$

then

$$\hat{\varepsilon} = \frac{c^2}{4\pi^2(l_a^2 - l_b^2)}[\varphi^2 - (\log_e S)^2 - 2i\varphi \log_e S] \qquad (3.79)$$

so that $\hat{\varepsilon}$ is now an explicit function of $H_a(f)$ and $H_b(f)$.

The main advantage of the transmission methods is that large phase changes are produced so that errors in time referencing are of less consequence than in the previous reflection method. Against this there is the disadvantage of a reduced frequency range of operation.

These problems have been considered by Loeb et al. (1971) and Suggett (1975), who suggest that the results from the transmission and reflection systems could be combined. Essentially the accurate phase information obtained from the transmission system is used to upgrade the data obtained from the reflection measurements. Thus high accuracy is obtained over a large frequency range.

Although many unwanted reflections can be eliminated with suitably placed air lines, the trigger pulse used to drive the tunnel diode can become superimposed on the observed waveform. This can then lead to difficulties in the Fourier transformation of the data. Suggett (1975) suggests putting a high-pass filter after the tunnel diode so that this low-frequency noise is eliminated, and claims that with such a system the errors in the time measurement may be as low as 0.08 ps.

(c) *The experimental cell*. An attractive feature of TDS is the simplicity of the experimental cell since no moving parts or measuring devices are required. Such a cell consists of a sample holder which enables a constant thickness of liquid to be introduced into a coaxial line.

A suitable cell has been described by Fellner-Feldegg (1969) and was constructed from an Amphenol APC-7 precision air line. Such lines have removable connectors and can easily be filled with liquid. When the connectors are replaced their precision is such that a liquid-tight seal is obtained. The cell used by Fellner-Feldegg (1969) and suitable for single reflection or transmission measurements is shown in Fig. 3.25. The Amphenol lines can be obtained in various lengths up to 30 cm. The size chosen must be a compromise between the volume of the test

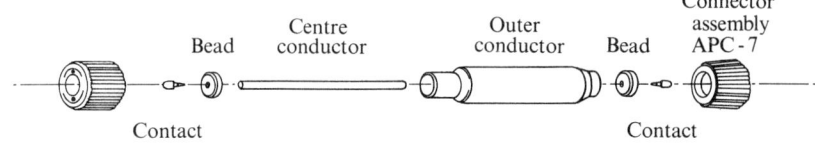

FIG. 3.25. A TDS cell suitable for single response methods (from Fellner-Feldegg 1972)

liquid used and the separation of the main and echo pulses, a length of 20 cm being useful for general purpose measurements.

(d) Analysis of the data and minimization of errors. In §§ 3.6.3(*a*) and 3.6.3(*b*) we have described basic TDS systems which although useful for mapping broad features of dispersions would, without additional equipment, be unlikely to yield high precision results. Problems are caused partly by the noise and jitter associated with a fast rise time pulse and partly by the difficulty in taking accurate measurements from an oscilloscope or chart recorder. To achieve a reasonable standard of accuracy, signal averaging and automatic data acquisition are required. This can be achieved with a commercial signal averager which will store the data in analog form, average it, and then punch the output on to computer tape.

An alternative and more flexible method is to use a small on-line computer. As well as the basic function of signal averaging this could, for example, produce a weighted average so that the most recently acquired data is given the greatest weight. It could also look for any long term instability within the system and possibly perform the Fourier transform on-line.

To consider the Fourier transform in more detail, the method employed by Loeb *et al.* (1971) involved the Samulon modification of the Shannon sampling theorem (Samulon 1951; Shannon 1949). This can be written as

$$R(f) = \frac{\Delta t}{1 - \exp(-2\pi i \Delta t)} \sum_{n=-\infty}^{\infty} \{r(n\Delta t) - r(n-1)\Delta t\} \exp(-2\pi i n \Delta t), \quad (3.80)$$

where Δt is the time interval between two samples. However, the use of eqn (3.80) can enhance the noise levels. This has been pointed out by Van Gemert (1973) who suggests subtracting a ramp voltage from the step response. Loeb (1972) has published a method for situations where a large number of time points are transformed into a few frequency points, whilst if a large number of frequency points are required then a fast Fourier transform may be more appropriate (Cooley and Tukey 1965).

To consider sources of error and their minimization, the need for signal averaging has already been mentioned. However, there may be long-term instability in the system. This can be detected and corrected by programming the system to keep re-measuring the first data point during the acquisition of a complete TDS trace.

Although the above method of data analysis by means of a Fourier transform could be considered as the correct approach, there is the disadvantage that the answers cannot be 'immediately seen'. Fellner-Feldegg (1969) attempted to derive frequency domain parameters directly from the time domain trace but, as pointed out in § 3.6.1 approximation of the parameter values is required. In a later paper (Fellner-Feldegg and Barnett 1970) it was shown that ε_s and ε_∞ can be obtained directly from the TDS trace and calibration graphs were provided to enable the relaxation time τ to be deduced. However, Van Gemert (1973) concludes that in a practical situation it is unlikely that both ε_s and ε_∞ could be directly deduced. Brehm and Stockmayer (1973) have also published calibration graphs.

Several theoretical papers have been published to calculate the pulse response of a system for various dielectric models (Van Gemert and de Graan 1972; Van Gemert and Bordewijk 1972; Bucci, Cortucci, Franceshetti, Savarese, and Tiberie 1972).

3.6.4. *Multiple response methods*

The single response methods of the previous section use experimental cells whose lengths are typically between 15 and 30 cm. With multiple response methods, however, the cell length is generally less than 1 mm. This is clearly an advantage for biological work since very little liquid is required, a typical cell volume being of the order of 1 μl.

(a) The exact method of analysis. The basic theoretical aspects of multiple response TDS have already been considered in § 3.6.2. The essence of the method is that the liquid is

contained in a very thin cell and the total signal is collected after being multiply reflected between the windows of the cell. Then, by means of Fourier transformation, the frequency domain response of the system can be obtained. Various techniques have been reviewed by Van Gemert (1973) and the theoretical aspects have been considered by Van Gemert (1974a) and Giese and Tiemann (1975). The method described below is due to Suggett and his co-workers (Clark *et al*. 1974).

The experimental system is basically that of Fig. 3.20. Suitable cells have been described by Fellner-Feldegg (1972) and are shown in Fig. 3.26. That shown in Fig. 3.26(a) has an inner conductor with a slightly reduced diameter at the position of the Teflon beads so that the line impedance is constant. The cell of Fig. 3.26(b) has an inner conductor of constant diameter. Teflon beads can be pressed on to this conductor and a removable spacer provides a well-defined gap for the test liquid. However, this cell has a small impedance mismatch at the beads so that an empty cell produces a reflection which has to be subtracted in the final analysis.

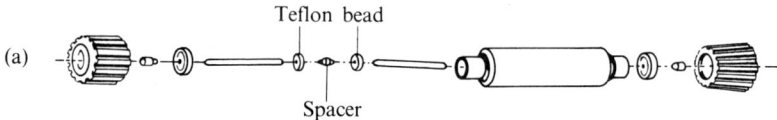

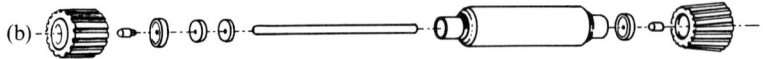

FIG. 3.26. TDS cells for multiple response methods: (a) a matched 50 Ω cell, (b) a small impedance mismatch (from Fellner-Feldegg 1972)

The filling procedure involves holding the line vertically and inserting the cell into the outer conductor until a gap for the liquid just remains. The sample is then injected with a syringe; the cell is pushed fully into the outer conductor and the connectors replaced, excess liquid being wiped off. For the purposes of stabilizing the temperature of the test liquid, a water jacket can easily be constructed around the cell.

Unlike the methods described in the next subsection, the total reflection system does not make any assumptions as regards the input pulse shape or the functional form of the dispersion relation for $\varepsilon(f)$. It has been suggested by Clark *et al.* (1974) that the cell length should be chosen such that the multiple reflections extend over about 50 per cent of the time window. For water or a peptide such as alanylglycine they used a cell of only 0.07 mm thickness. For a large globular protein such as myoglobin a relatively thick cell was used, but this was still only 5 mm long.

The experiment involves recording the totally reflected signals from the sample and then from a short circuit placed in the same position as the sample. Typical waveforms are shown in Fig. 3.27. The scattering coefficient $S_{11}(f)$ of eqn (3.76) can then be obtained from the ratio of the Fourier transforms of the two signals. By means of eqns (3.73) and (3.67), $\varepsilon(f)$ can be obtained from eqn (3.76) by numerical methods, such as the Newton-Raphson, as explained in § 3.6.2.

Clearly the above method could be extended to longer times by using longer cells, i.e. observing multiple reflections from a long cell. However, Clark (Clark *et al.* 1974) experienced difficulty in establishing the position of the top of the pulse before reflection and suggested the method outlined in Fig. 3.28. This involves (a) the incident pulse, and (b) the incident plus the pulse reflected from the sample (Figs. 3.28(a) and (b)). The signal of (b) is then subtracted from (a) to give the waveform shown in Fig. 3.28(c). The ratio of the Fourier transforms of traces (c) and (a) then gives the scattering coefficient.

(b) Approximate methods of analysis. The thin cell techniques are well suited to approximate methods of analysis. Fellner-Feldegg (1972) proposed a thin cell method in which the scattering coefficients were expanded into a Taylor series, only the first-order terms being retained. This technique enabled $\hat{\varepsilon}$ to be obtained explicitly. The method has also been considered in more detail by Van Gemert (1973, 1974*a*). However, all such techniques are an approximation since assumptions have to be made regarding the initial pulse shape and the functional

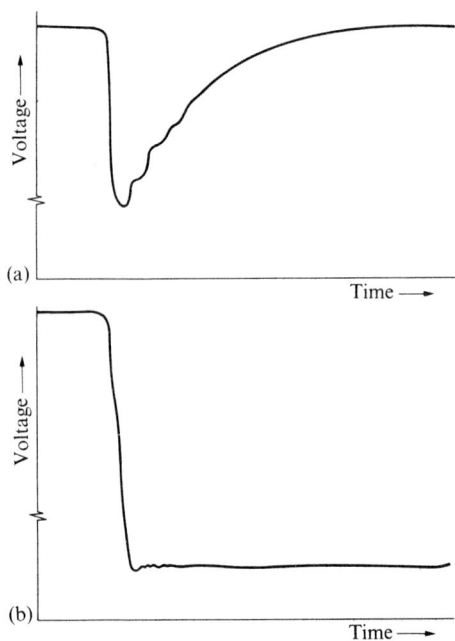

FIG. 3.27. Typical waveforms obtained from multiple response TDS: (a) the signal reflected from a sample, (b) the signal from a short circuit

form of $\varepsilon(f)$. Restrictions also have to be placed on the size of the cell and the relaxation parameters of the sample.

Most methods assume $2\alpha l \ll 1$ so that $\exp(-2\alpha l)$ can be approximated to $1 - 2\alpha l$. Van Gemert (1974a) has shown that for a single Debye-type process the time-dependent reflection coefficient can be written as

$$r(t) = \frac{-l}{2c\tau}(\varepsilon_s - \varepsilon_\infty) \exp(-t/\tau), \qquad (3.81)$$

where the symbols have their usual meaning. However, the sample length and time scales are restricted according to

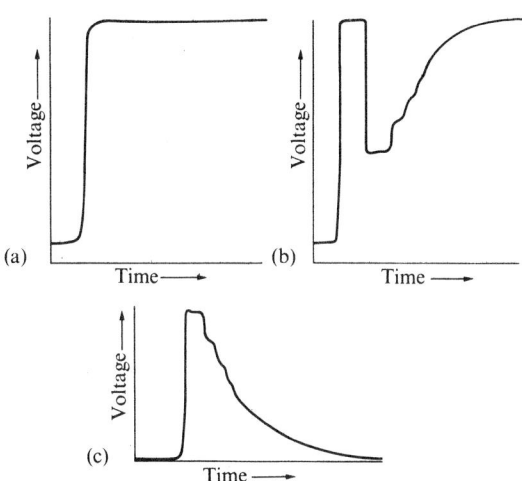

FIG. 3.28. An alternative TDS method: (a) the incident pulse, (b) the incident + the signal reflected from the sample, (c) the effect of subtracting (b) from (a) (from Clark et al. 1974)

$$(l/2c\tau)(\varepsilon_s - \varepsilon_\infty)t/\tau \ll 1 \tag{3.82}$$

$$(l/2c)|\tau\varepsilon_s - 3\varepsilon_\infty| \ll 1. \tag{3.83}$$

Eqn (3.81) can be generalized to overlapping Debye processes, giving

$$r(t) = \frac{-l}{2c}\left\{\frac{\Delta_1}{\tau_1}\exp(-t/\tau_1) + \frac{\Delta_2}{\tau_2}\exp(-t/\tau_2) + \ldots\right\}, \tag{3.84}$$

where Δ_i is the dielectric increment of the ith component.

The practical aspects have been considered by Clark et al. (1974). To minimize the effect of unwanted reflections from the cell windows, which in practice cannot be perfectly machined, the signal reflected from the dielectric filled cell is subtracted from that reflected from an ampty cell as shown in Fig. 3.29. Eqns (3.81) and (3.84) assume that the input pulse is perfect, i.e. the rise time must be very much less than the

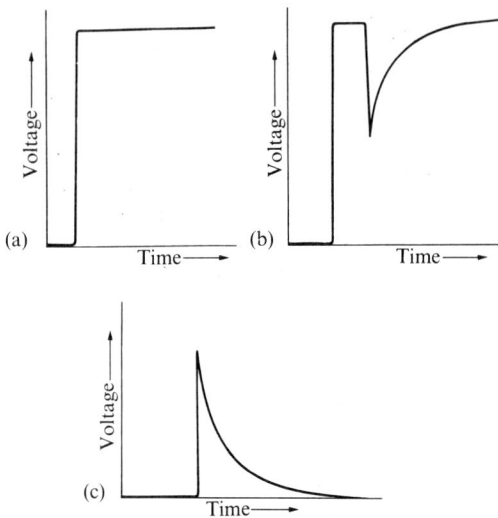

FIG. 3.29. TDS traces obtained from the thin cell method: (a) the incident pulse, (b) the pulse from the liquid-filled cell, (c) the effect of subtracting (b) from (a) (from Clark et al. 1974)

time-scale of measurement. In practice this restricts the method to measurements at times greater than 50 ps.

Methods of analysis have also been considered recently by Cole (Cole 1974, 1975a, b, c). These papers contain equations relating permittivity to observed TDS traces and will under certain circumstances reduce to the equations derived by Fellner-Feldegg and Van Gemert. However, the analysis is more general since assumptions do not have to be made regarding the functional form of $\varepsilon(f)$. Also corrections are given for the finite rise time of the initial pulse. Readers wishing to use such methods rather than the 'exact analysis' of § 3.6.4(a) are strongly advised to consult these papers.

3.6.5. *Conductive solutions*

Like the frequency domain methods discussed earlier in this chapter the presence of conductivity will make TDS measurements difficult and inaccurate and a high conductivity may make

permittivity measurements impossible. Van Gemert (1973) has
considered this problem and concludes that with the direct
reflection method it should in theory be possible to obtain the
conductivity σ from the differential of the $v(t)$ curve at $t = 0$.
However, in practice experimental error would be too large and
ε_s, ε_∞, and τ would have to be known. Although in theory σ
could be obtained after the Fourier analysis of the decay
curve, this decay will be slowed down by conductivity so that
for small or medium conductivities the time window will be too
small. For large conductivities the decay is fast enough for
Fourier analysis to yield values of σ but under these conditions it would then not be possible to obtain the permittivity
$\hat{\varepsilon}$. Progress is being made towards solving these problems
(Clarkson, Glasser, Tuxworth, and Williams 1977) and we have
recently measured to within ±1 per cent the permittivity of a
solution of ATP of concentration 14 per cent, which has a conductivity of 1.5×10^{-2} Ω^{-1} cm^{-1} (Dawkins, Sheppard, and Grant
1978).

However, with the thin cell methods the determination of
conductivity is possible. It has been shown (Fellner-Feldegg
1972; Van Gemert 1973) that the TDS trace shows an offset in
the base line which is a function of σ. Thus with such methods
it is possible to calculate σ independently of the dipolar
relaxation.

3.6.6. *Conclusions on time domain measurements*

The previous sections have outlined various TDS systems, but
other alternatives exist. For example, Nicolson and Ross (1970)
have an experimental method similar to that of Fig. 3.20 but a
short circuit is placed at a distance d beyond the thin cell.
With this system both the totally reflected and totally transmitted signals will be seen, and provided that d is large
enough they will be separated in time. The results are then
Fourier analysed and Nicolson and Ross show that $\hat{\varepsilon}$ can be obtained explicitly from the observed scattering coefficients.

A novel method has been proposed by Iskander and Stuchly
(1972) and by Rzepecka and Stuchly (1975) where the dielectric
material is not placed between the inner and outer conductors.
Instead the line is terminated with a short circuit but the

inner conductor does not quite make contact with it. The sample is then placed between the end of the inner conductor and the short circuit. This system can be considered theoretically as a shunt capacitor located at the end of the transmission line (Van Gemert 1973, 1974*b*).

For very low-frequency work the method of Hyde (1970) is a modern version of the Davidson *et al.* (1951) system whereby the effect of charging or discharging a condenser is observed. This method is particularly suited to low-frequency measurements and is novel in taking results at logarithmically spaced times.

Since TDS is a technique which has only been in existence for a decade it is difficult at this stage (1978) to advise newcomers to the field of dielectric measurements whether to opt for TDS rather than for the more traditional frequency domain methods. This is particularly so when biological liquids are being considered. Unfortunately in the recent past some workers have been rather over-enthusiastic in claiming that permittivity could be measured using time domain methods with great savings of time, with little experimental expertise, and with a minimum of capital outlay. Certainly it is now realized that with the necessary signal averaging equipment the cost of TDS can approach that of a frequency domain system and high accuracy can only be obtained by a skilled operator. Moreover, the overall saving in the time taken to complete a run of measurements is debatable. It is thus felt that in the present state of the art, frequency domain methods are more satisfactory when the highest accuracy in permittivity is required or when small dispersions are to be resolved. On the other hand, for situations where well-defined dispersions are to be studied, then TDS is to be recommended. At present, when resources permit, the ideal situation is to have both frequency domain and TDS apparatus since one can complement the other. Fortunately research into TDS is progressing quite rapidly and possibly in the not too distant future it may come to be regarded as a suitable method for all types of dielectric investigation.

4
THE ANALYSIS OF EXPERIMENTAL RESULTS

4.1. General introduction

The previous chapter described methods of measuring complex permittivity. Before such data can be interpreted in terms of molecular structure an underlying theoretical model has to be postulated. Perhaps it should be mentioned that the term 'model' has a rather different meaning in this chapter. In previous usage, 'model' has referred to the shape or structure of a physical system, such as the configuration of atoms in a particular protein molecule. However, in this context 'model' is being used in the statistical sense to indicate a mathematical expression which a set of experimental data may tend to follow. For example, the model may be the equation of a single Debye dispersion, or a distribution of relaxation times, or possibly two single Debye terms. Once the most suitable model has been chosen, then its parameters can be calculated and a molecular interpretation attempted.

Until a decade ago this determination of the model and its parameters was achieved mainly by graphical methods. Although most of these methods have been superseded by computer techniques some are still useful for suggesting the appropriate model for the computer analysis. For this reason the most important graphical methods are briefly summarized in § 4.2.

However, the main objective of this chapter is to explain the computer methods of data analysis with particular reference to least-squares curve fitting. Unfortunately dangers do exist and for this reason much emphasis is placed upon the statistical treatment of the data in order that the validity of conclusions drawn can be rigorously assessed.

4.2. Graphical methods for preliminary analysis of data

The basic Debye equations have been introduced in Chapter 2 where eqns (2.43), (2.44), (2.45), and (2.46) are derived. They are shown graphically in Figs. 2.12 and 2.13. If the experimental dispersion ε' is not well represented by a single relaxation time then it will cover a wider range of frequencies than required by the above equations, appearing broader in shape with a more shallow slope at the centre. The curves for ε'' or s will also be broader.

The possibility of a model which is more complicated than a single Debye dispersion has already been considered in § 2.3, where a distribution of relaxation times was expressed as the convolution of a Debye function with a distribution function (eqn (2.70)). This approach is considered further in § 4.6; however, the numerical analysis involved is not trivial and simpler graphical methods are initially considered.

For example, from eqn (2.44)

$$\frac{\varepsilon' - \varepsilon_\infty}{\varepsilon_s - \varepsilon'} = \frac{1}{\omega^2 \tau^2} , \qquad (4.1)$$

so that a plot of $(\varepsilon'-\varepsilon_\infty)/(\varepsilon_\infty-\varepsilon')$ against $1/\omega^2$ gives a straight line from whose slope the relaxation time τ may be calculated. If, however, the points do not lie on a straight line, then they are not represented by a single relaxation time. Although this method only requires ε' data, there is the practical problem that measurements are required throughout the entire dispersion region and this is not always possible.

Another useful plot is that of s against ε'. The elimination of ω from eqns (2.44) and (2.46) gives

$$\varepsilon_s - \varepsilon' = \tau(s - s_s) , \qquad (4.2)$$

where s_s is the limiting value of s at low frequencies. Thus for a single relaxation time, a graph of ε' against s gives a straight line of slope $-\tau$. Although data need not be taken over the complete dispersion, the limiting values of ε' and s at low frequences, namely ε_s and s_s are required and these can at times be difficult to measure.

An alternative graphical method, introduced by Cole and Cole (1941), overcomes the problems associated with measuring data over a large frequency range or at the low frequency end of a dispersion. If ε'' is plotted against ε', then for a single Debye dispersion a semi-circle results with its centre on the ε' axis (Fig. 4.1). If, however, a distribution of relaxation times exists, then the semi-circle will have its centre depressed below the ε' axis. With this method a semi-circle and the position of its centre can often be recognized from an arc rather than a complete semi-circle so that data are not required over the whole dispersion range. The empirical frequency domain equation which leads to the Cole–Cole circle is

$$\hat{\varepsilon} = \varepsilon_\infty + \frac{\varepsilon_s - \varepsilon_\infty}{1 + (i\omega\tau)^{(1-\alpha)}}, \qquad (4.3)$$

where α is a measure of the spread of the relaxation times (known as the Cole–Cole α) whose magnitude can be estimated from the Cole–Cole plot (see Fig. 4.1). From eqn (4.3) it can be seen that when $\alpha = 0$ a single Debye dispersion equation (2.43) results. Other graphical methods of analysis based on eqn (4.3) have been derived by Grant (1969), Grant, Buchanan, and Cook (1957), and Williams (1959).

The best estimates of dispersion parameters, particularly when all of the data points are not equally accurate, can be obtained only with a computer analysis. Also such methods enable parameter errors to be rigorously estimated by means of

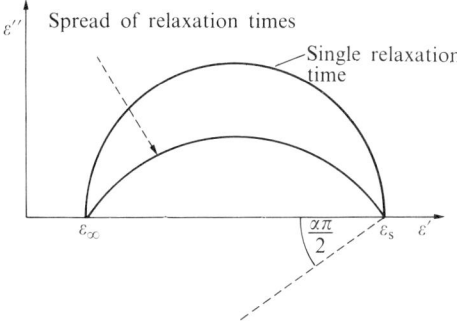

FIG. 4.1. The Cole–Cole plot

a statistical analysis. Thus although the above graphical techniques are still useful for a pre-computer visual inspection of the data, they should not be used in any final data analysis. Because the computer is now such a valuable tool in the field of dielectric measurements the rest of this chapter concentrates entirely on computerized techniques for data handling. Initially least-square analysis and selected statistical methods are considered in general terms. However, the more practical questions regarding the possible models to be fitted and in particular the choice of the best model are considered in §§ 4.4.5, 4.5, and 4.6.

4.3. Least-squares minimization
4.3.1. *Introduction*

In this section it will be assumed that the best model to be fitted is known and that only its parameters are required. Problems associated with the correct choice of model and the validity of its parameters will be discussed later.

To consider a practical example, suppose that the real part of the permittivity, ε', is to be fitted to the equation of a single Debye dispersion, i.e. eqn (2.44).

This can in general be considered as fitting a variable $y(\varepsilon')$ to $x(\omega)$ with parameters θ_1, θ_2, and θ_3 (ε_∞, ε_s, and τ respectively in eqn (2.44)). (The dielectric increment Δ referred to in eqn (2.44) is equal to $\varepsilon_s - \varepsilon_\infty$.) Thus for n parameters eqn (2.44) could be written in general terms as

$$y = f(x, \theta_1, \theta_2, \ldots, \theta_n), \qquad (4.4a)$$

or using vector notation

$$y = f(x, \boldsymbol{\theta}), \qquad (4.4b)$$

where $\boldsymbol{\theta} = (\theta_1, \theta_2, \ldots, \theta_n)$.

Experimental values of y (say Y_i) are measured at various values of x, x_i ($i = 1, 2, \ldots, m$), but since the Y_is are subject to experimental error, eqn (4.4b) becomes

$$Y_i = f(x_i, \theta) + \varepsilon_i, \qquad (4.4c)$$

where ε_i is the error on the ith experimental point. It is unfortunate that ε_i is conventionally used to denote error and care is required not to confuse it with permittivity $\hat{e}$. For all of the m experimental points the sum of squares of the errors is given by

$$\Phi = \sum_{i=1}^{m} \varepsilon_i^2 = \sum_{i=1}^{m} [Y_i \ f(x_i, \theta)]^2 \ . \qquad (4.5)$$

The purpose of least-squares minimization is to find values of the parameters $\theta = (\theta_1, \theta_2, \ldots, \theta_n)$ which make Φ a minimum.

It should be noticed that eqn (4.4c) assumes that x_i is measured without error or with negligible error. This is, however, valid for most dielectrics applications; in eqn (2.44), for example, the error in ε' is normally much greater than the error in the angular frequency ω.

When the model is linear in the parameters, the best fit values of θ can easily be calculated (Draper and Smith 1966). However, for most applications in dielectrics the models are non-linear. For such cases there is no exact solution for θ and an iterative technique has to be used. For example, the parameters can be estimated and then each one successively varied until values are found which make Φ a minimum. The most satisfactory of these systematic trial and error methods is the Simplex (Spendley, Hext, and Himsworth 1962; Nelder and Mead 1965). This method is particularly good for difficult functions but tends to be slow to converge. A rather more mathematical method (Rosenbrock 1960) estimates the 'direction' of the minimum by evaluating the differentials of Φ with respect to each parameter, i.e. $\partial\Phi/\partial\theta_j$. This method converges more rapidly than those of the Simplex type but requires a function whose differentials can be calculated.

For dielectric applications we have found the Taylor expansion or Marquardt method to be the most suitable (Marquardt, Bennett, and Burrell 1961; Marquardt 1963). This procedure also evaluates $\partial\Phi/\partial\theta_j$ but at each iteration the corrections to the parameter estimates are expressed as a simple matrix equation which is ideal for computational purposes. The method is fast to converge and although continuous well-behaved functions are required, these are usually encountered in this kind of work.

4.3.2. The Marquardt method

(a) The Marquardt method for real functions. The theory of this method involves expanding the function to be fitted (eqn (4.4b)) as a Taylor series of which only the first-order differential terms are retained. For m experimental points this leads to a system of m equations whose solution is given by

$$\Delta\boldsymbol{\theta} = (Z'Z)^{-1}Z'\Delta\mathbf{Y}, \qquad (4.6)$$

where

$$\Delta\boldsymbol{\theta} = \begin{pmatrix} \Delta\theta_1 \\ \vdots \\ \Delta\theta_n \end{pmatrix}$$

is a correction to the previous estimated value of $\boldsymbol{\theta}$. Z is an $(m \times n)$ matrix given by

$$Z = \begin{pmatrix} \dfrac{\partial f_1}{\partial \theta_1} & \cdots & \dfrac{\partial f_1}{\partial \theta_n} \\ & \dfrac{\partial f_i}{\partial \theta_j} & \\ \dfrac{\partial f_m}{\partial \theta_1} & \cdots & \dfrac{\partial f_m}{\partial \theta_n} \end{pmatrix}$$

and $\partial f_i/\partial \theta_j$ denotes the function (eqn (4.4b)) differentiated with respect to the jth parameter and then evaluated at the ith experimental point. Z' denotes the matrix transposition and Z^{-1} the matrix inversion of Z respectively, whilst $\Delta\mathbf{Y}$ is the vector

$$\Delta\mathbf{Y} = \begin{pmatrix} Y_1 - y_1 \\ \vdots \\ Y_m - y_m \end{pmatrix},$$

i.e. the difference between the m experimental and calculated points. The derivation of eqn (4.6) will not be given since it is well documented in the literature (Marquardt et al. 1961; Marquardt 1963).

After each iteration the corrected values of the parameters are used in the calculation of a root mean square error defined by

$$\text{RMSE} = \sqrt{\frac{\sum_{i=1}^{m}[Y_i - f(x_i, \theta)]^2}{m}}. \qquad (4.7)$$

When two successive values of RMSE agree to within a small number (usually of the order of 10^{-6}) then the fitting is terminated and the last values of θ are taken to be the best fit values of the parameters.

It can be seen that eqn (4.7) is not normalized so that the limit of convergence being set at 10^{-6} may appear arbitrary. However, this is satisfactory in practice provided that the data are kept to within a range of 10^{-3} to 10^{+3}. For example, a frequency of 10 000 MHz would be programmed as 10 GHz and *not* 10^{10} Hz.

So far nothing has been said about the nature of the error term ε_i. For statistical purposes (see § 4.4) certain conditions such as independence and normality of distribution for ε_i are important (see Draper and Smith 1966). These assumptions are usually valid for dielectrics data. However, one important condition which may not be fulfilled is that the expected variances of the ε_i should be constant. Less precisely, we can say that the experimental errors on each of the data points should be approximately equal. If some points have larger errors then they should be given less weight in the fitting procedure. This can be expressed as follows. If each experimental point Y_i has an error ε_i with a variance of σ_i^2, then we can define a weight matrix as

$$W = \begin{pmatrix} \frac{1}{\sigma_1^2} & & & 0 \\ & \ddots & & \\ & & \frac{1}{\sigma_i^2} & \\ & & & \ddots \\ 0 & & & \frac{1}{\sigma_m^2} \end{pmatrix}. \qquad (4.8)$$

The determination of σ_i^2 will be considered in §4.6. The off-diagonal elements of W are zero; this expresses the fact that the observations are independent. With the inclusion of weighting factors eqn (4.6) becomes

$$\Delta\theta = (Z'WZ)^{-1} Z'W\Delta Y. \qquad (4.9)$$

When all of the data points are equally accurate each diagonal element of W is 1 and eqn (4.9) reduces to eqn (4.6).

(b) *The Marquardt method for complex functions.* In many dielectric applications the function to be fitted is complex, for example the equation of a single Debye dispersion has been given (eqn (2.43)) and can be written as

$$\hat{\varepsilon} = \varepsilon_\infty + \frac{\varepsilon_s - \varepsilon_\infty}{1 + 2\pi i f \tau}, \qquad (4.10)$$

where $\hat{\varepsilon}$ is a complex number of the form $\varepsilon' - i\varepsilon''$ and f is the frequency. Although ε' and ε'' could be fitted separately problems arise since each fit will in general give a different estimate for each parameter. These separate estimates must then be combined with a *weighted* average but the appropriate weighting factors *cannot* be deduced from the data.

It can be shown (Sheppard, Jordan, and Grant 1970) that the full complex data can be used to give parameter estimates of

$$\Delta\theta = (Z_R'Z_R + Z_x'Z_x)^{-1}(Z_R'\Delta Y_R + Z_I'\Delta Y_I), \qquad (4.11)$$

where Z_R denotes the real and Z_I the imaginary part of the now complex matrix Z; all other symbols should be clear from the previous section.

For weighted data there is now a complex weight matrix of the form $W = W_R + iW_I$ where the jth diagonal element of W is given by

$$W_{jj} = \frac{1}{\sigma_{R_j}^2} + i\frac{1}{\sigma_{I_j}^2} \qquad (4.12)$$

and $\sigma_{R_j}^2$ and $\sigma_{x_j}^2$ are the variances of the real and imaginary parts respectively of the jth complex data point. The estimate of $\Delta\theta$ is now given by

$$\Delta\theta = [Z_R'W_R Z_R + Z_I'W_I Z_I]^{-1} \cdot [Z_R'W_R \Delta\mathbf{Y}_R + Z_I'W_I \Delta\mathbf{Y}_I]. \qquad (4.13)$$

4.3.3. *Practical details for a curve-fitting program*

The differentials of the matrix Z (eqn (4.6)) for a real function or eqn (4.11) for a complex function can be evaluated numerically using

$$\frac{\partial f_i}{\partial \theta_j} \simeq \frac{f(x_i, \theta_1, \ldots, \theta_j + \Delta\theta_j, \ldots, \theta_n) - f(x_i, \theta_1, \ldots, \theta_j, \ldots, \theta_n)}{\Delta\theta_j}, \qquad (4.14)$$

where $\Delta\theta_j$ is a small number. Thus for a general fitting program only the function need be defined, the differential terms of Z being calculated by the computer.

Such a general fitting program needs certain additional features to deal with difficult data. It is often advisable to initially fit only some of the parameters, generally the most important ones, and to leave the others until an approximate convergence has been obtained. As an example consider the standing wave in a coaxial line (eqn (3.49)), where the parameters ρ and x_0 would be held at 1 and 0 respectively whilst ε', ε'', and db_0 were fitted. When two successive values of the RMSE agreed to within 0.1 then all five parameters would be varied.

Due to the approximations used in the theory of the Marquardt method, the calculated value of $\Delta\theta$ will sometimes overcorrect a parameter and possibly cause divergence, i.e. the RMSE increases. In such cases $\Delta\theta$ is multiplied by a factor F, where F starts at unity and is successively divided by two. If this restores convergence, then at each future iteration F is multiplied by two until it either attains a value of unity or divergence occurs again, so that F is again reduced. If F gets smaller than say 2^{-20} and convergence is not obtained, then the minimization is terminated and a warning message is printed. However, with dielectric data this rarely occurs unless there are high correlation coefficients between some parameters due to too many parameters being fitted (see § 4.4.3) or when overlapping dispersion regions are involved (§ 5.5).

Sometimes in the early stages of a minimization a parameter will go far away from its best-fit value and then not be able to return. This can be prevented if upper and lower limits are put on to each parameter. However, if after a minimization any parameter has 'stuck' on a limit, then the data *must* be run again with wider limits.

A listing of a suitable least-squares curve-fitting program has not been given since this should be available in any good computer library. In most cases it is only necessary to supply the appropriate subroutine in order to define the function actually being fitted.

4.4. The statistical analysis of data

4.4.1. *Introduction*

The previous section showed how the parameters of a model may be estimated from experimental data. However, before valid molecular conclusions can be drawn it is necessary to consider the accuracy and reliability of both the model and its parameters. The objective of this section is to demonstrate how this can be done with the aid of a simple statistical analysis.

4.4.2. *Confidence intervals and the t-distribution*

Initially it will be assumed that the model to be fitted is known and that only the accuracy of its parameters is in doubt. When data are being analysed by the method of least squares it is preferable to calculate a confidence interval for each parameter. From eqn (4.9) the variance–covariance matrix of the parameters M is defined as

THE ANALYSIS OF EXPERIMENTAL RESULTS   131

$$M = (Z'WZ)^{-1} \tag{4.15}$$

where for unweighted data all the diagonal elements of W will be equal to one. For the jth parameter its confidence interval is

$$\Delta\theta_j = \sqrt{M_{jj}} \sigma t_{(m-n)}, \tag{4.16a}$$

i.e. the parameter and its confidence interval c is

$$c = \theta_j \pm \Delta\theta_j, \tag{4.16b}$$

where M_{jj} is the jth diagonal element of the matrix M and σ and $t_{(m-n)}$ will be explained below. The overall variance of the fit, σ^2, is defined as

$$\sigma^2 = \sum_{i=1}^{m} \frac{[Y_i - y_i]^2}{(m-n)}, \tag{4.17}$$

where Y_i is the experimental and y_i the calculated value of y.

Thus from eqn (4.7)

$$\sigma^2 = \frac{m}{(m-n)} (RMSE)^2, \tag{4.18}$$

where m is the number of experimental points and n the number of parameters. The factor $t_{(m-n)}$ is the appropriate percentage point of the t-distribution with $(m-n)$ degrees of freedom (Draper and Smith 1966, Brownlee 1965).

To consider the above in more detail; confidence intervals are quoted at a given probability level, 95 per cent and 99 per cent being the most usual. A 95 per cent confidence interval means that there is a 95 per cent chance of the interval $(\theta_j + \Delta\theta_j)$ to $(\theta_j + \Delta\theta_j)$ covering the true value of the jth parameter. The value of t also depends upon a parameter v known as the degree of freedom; in the above case $v = (m-n)$. The t-distribution is tabulated in statistical tables (see, for example, Lindley and Miller 1964). An extract from such a set of tables is given in Table 4.1. The method of tabulation for

TABLE 4.1

Percentage points of the t-*distribution*
(from Lindley and Miller 1964)

Degrees of freedom	Percentage points for 95 per cent confidence interval	Percentage points for 99 per cent confidence interval
1	12.71	63.66
2	4.30	9.92
3	3.18	5.84
4	2.78	4.60
5	2.57	4.03
6	2.45	3.71
7	2.36	3.50
8	2.31	3.36
9	2.26	3.25
10	2.23	3.17
12	2.18	3.05
15	2.13	2.95
20	2.09	2.85
24	2.06	2.80
30	2.04	2.75
40	2.02	2.70
60	2.0	2.66
120	1.98	2.62
∞	1.96	2.58

the t-distribution can vary; often the percentage points for a 95 per cent confidence interval are called 5 per cent or even 2½ per cent points. However, the form of tabulation of an unknown set of tables should become clear when compared with Table 4.1. This table shows that as the degrees of freedom increase, the value of t decreases and hence demonstrates the desirability of a large number of experimental points. For $v > 10$ the value of t is in the order of two. An error of plus or minus one standard deviation (i.e. a t value of one) corresponds to a confidence interval of about 66 per cent. Although ±1 standard deviation has sometimes been quoted in the literature, most errors have not been calculated on a rigorous basis. Thus it is difficult to compare the accuracy of modern work, estimated on the basis of a 95 per cent confidence interval, with the errors quoted in much of the previous literature.

It is important that the limitations of a confidence interval are clearly stated. Such intervals cannot give an absolute indication of the total error on a parameter. An assumption is made that the errors are random and normally distributed. Although the assumption of normality is generally valid, it is vital to notice that any systematic *errors* will *not* be included in a confidence interval. Also such an interval is valid only if the model being fitted is correct.

4.4.3. *Correlation coefficients*

Ideally the parameters of any model should be independent. However, in a practical situation correlations will always exist. The correlation coefficient ρ_{ij} between the parameters i and j can be calculated by

$$\rho_{ij} = \frac{M_{ij}}{\sqrt{M_{ii} M_{jj}}}, \qquad (4.19)$$

where M_{ij} is the ij^{th} element of the variance–covariance matrix M (eqn (4.15)). If the parameters i and j are independent then $\rho_{ij} = 0$, whilst $\rho = 1$ or -1 denotes perfect positive or negative correlation.

After a least-squares minimization the correlation coefficients between all parameters should be calculated. Although ideally all such coefficients should be zero, this will not occur in practice. The calculated correlation coefficient is a practical rather than theoretical value. To explain this with an example, consider fitting ε' data to a single Debye function (eqn (2.44)). In many practical situations only low-frequency data are available and may appear as in Fig. 4.2. Although this example has been deliberately exaggerated it demonstrates the point being made. It can be seen that it is difficult to obtain τ (or the relaxation frequency f_R) and ε_∞ from such data when experimental error is present. Although

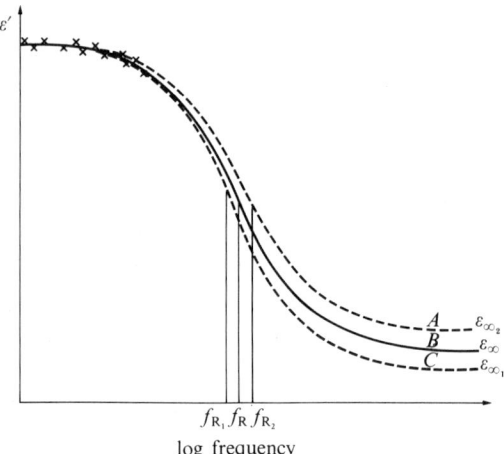

FIG. 4.2. Practical correlation between f_R and ε_∞. Data points are shown as crosses. The solid curve is the best fit line and the dashed curves are reasonable alternative lines

curve B of Fig. 4.2 may be the best fit, curves A and C will give very little variation in the root-mean-square error of fit at the *observed* data points. Thus either parameters f_{R_1} and ε_{∞_1} or f_{R_2} and ε_{∞_2} give equally good fits to the data so that, although physically independent, f_R and ε_∞ appear to be correlated. This practical correlation coefficient is, however, of great value since if close to +1 or -1 it will indicate that either the variables are physically correlated, that too many parameters are being fitted to insufficient data, or that parameters insensitive to the experimental data are being fitted. Thus a large correlation coefficient acts essentially as a warning signal that the parameter values concerned may be unreliable.

4.4.4. *Confidence contours*

§ 4.4.2 has shown how confidence intervals may be put upon individual parameters. However, this procedure can rightly be criticized on the grounds of it being unreasonable to consider correlated parameters in isolation (i.e. $\rho_{ij} \neq 0$ for all i and j). This objection can be overcome by calculating a confidence contour rather than individual confidence intervals.

THE ANALYSIS OF EXPERIMENTAL RESULTS

For a model with two parameters θ_1 and θ_2 a confidence contour can be calculated by considering pairs of values of θ_1 and θ_2 for which the following equation is true:

$$\Phi = \Phi_{min}[1 + \frac{n}{m-n} F_{(n,m-n)}], \qquad (4.20)$$

where m is the number of experimental points and n the number of parameters (2 in this example). The sum of squares of the errors is denoted by Φ_{min} at the best-fit values of the parameters and by Φ at any other point. The factor $F_{(n,m-n)}$ is the appropriate percentage point of the F distribution with $(n,m-n)$ degrees of freedom; this distribution will be considered further in the next section. For more details of eqn (4.20), see Draper and Smith (1966). The values of θ_1 and θ_2 which satisfy eqn (4.20) will define a confidence contour. If θ_1 and θ_2 are linear and uncorrelated then this contour will be an ellipse, as shown in Fig. 4.3(a). If, however, individual intervals were calculated from eqns (4.16), then the contour

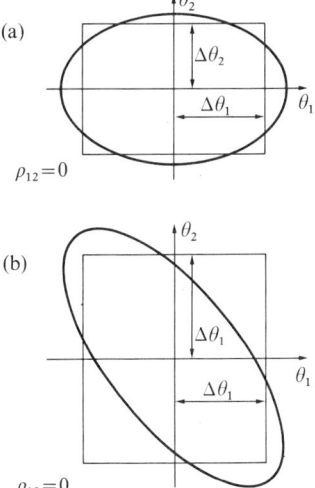

FIG. 4.3. Confidence contours for a two-parameter linear model: (a) the parameters θ_1 and θ_2 are not correlated, (b) the parameters θ_1 and θ_2 are correlated

would be a rectangle as shown in Fig. 4.3(a) and it can be seen that the two regions are similar. When θ_1 and θ_2 are correlated the ellipse will be rotated with respect to the θ_1, θ_2 axes so that the two confidence regions are now quite different (see Fig. 4.3(b)). Thus when parameters are highly correlated not only will the estimates of them be suspect, their confidence intervals will be unreliable.

For the more practical situation of a model which is nonlinear in the parameters, eqn (4.20) can still be solved to give values of θ_1 and θ_2 but the confidence contour will not be an ellipse and the probability level may be incorrect. Nevertheless, this method does give a useful approximate estimation of error whilst the amount by which the shape of the confidence contour departs from that of an ellipse gives an indication of the lack of linearity in the model.

The above discussion suggests that confidence contours rather than confidence intervals should always be calculated. However, problems arise when the model has more than two parameters. For an n-parameter system the confidence contour would be a region in n-dimensional space. The co-ordinates of this could still be calculated from eqn (4.20), with a computer, but the problems of interpretation are almost insoluble. The main objective of a confidence interval is to give a quick and easy indication of the accuracy of parameters and clearly it is difficult to visualize a region in n-dimensional space. Thus for most practical situations confidence intervals are the most suitable form of error estimation provided that their limitations are realized, particularly when correlated parameters are encountered.

4.4.5. *Choosing the best model—the* F-*distribution*

So far we have only considered the problems associated with determining the parameters of a model, assuming that the correct model is known. But frequently this is not the case. To consider a practical example, the permittivity of a substance, $\hat{\varepsilon}$, may be known as a function of frequency and it may be suspected that the underlying model is of a Debye type. However, it is often not obvious from inspection of the data whether the best model is a single or possibly the sum of several Debye dispersions.

Formally this problem can be considered as follows. There are two models, one with p and the other with n parameters where $p > n$. It is required to test whether the extra $(p-n)$ parameters significantly improve the fit. To digress for a moment; in general an increase in the number of parameters will reduce the RMSE. The purpose of the statistical test is to discover whether a significant improvement has been obtained, i.e. whether we can conclude that the model of p parameters is the better model. The following sums of squares are calculated:

$$\Phi_p = \sum_{i=1}^{m} (Y_i - y_{p_i})^2 \quad (4.21)$$

$$\Phi_n = \sum_{i=1}^{m} (Y_i - y_{n_i})^2, \quad (4.22)$$

where Y_i is the ith observed and y_{p_i} the ith calculated value for the model with p parameters. The value of H is then calculated where

$$H = \frac{(\Phi_n - \Phi_p)/(p-n)}{\Phi_p/(m-n)} \quad (4.23)$$

(see Draper and Smith (1966) for further details).

For the statistical test the F-distribution is required. This is tabulated (Lindley and Miller (1964)) for various percentage points, but unlike the t-distribution has two sets of degrees of freedom v_1 and v_2. An example of this distribution is given in Table 4.2 for a 95 per cent test. For the above case the degrees of freedom are $(p-n, m-n)$, i.e. $v_1 = (p-n)$ and $v_2 = (m-n)$ (see Table 4.2). If the value of H calculated from eqn (4.23) is greater than the tabulated F-value then we conclude that the increase in the number of parameters from n to p appears to be justified.

4.5. Practical considerations

The purpose of this section is to show how the statistical concepts of § 4.4 may be used in the analysis of permittivity

TABLE 4.2

5 per cent points of the F-distribution (for a 95 per cent test)
(from Lindley and Miller 1964)

v_2 \ v_1 =	1	2	3	4	5	6	7	8	10	12	24	∞
1	161.4	199.5	215.7	224.6	230.2	234.0	236.8	238.9	241.9	243.9	249.0	254.3
2	18.5	19.0	19.2	19.2	19.3	19.3	19.4	19.4	19.4	19.4	19.5	19.5
3	10.13	9.55	9.28	9.12	9.01	8.94	8.89	8.85	8.79	8.74	8.64	8.53
4	7.71	6.94	6.59	6.39	6.26	6.16	6.09	6.04	5.96	5.91	5.77	5.62
5	6.61	5.79	5.41	5.19	5.05	4.95	4.88	4.82	4.74	4.68	4.53	4.36
6	5.99	5.14	4.76	4.53	4.39	4.28	4.21	4.15	4.06	4.00	3.84	3.67
7	5.59	4.74	4.35	4.12	3.97	3.87	3.79	3.73	3.64	3.57	3.41	3.23
8	5.32	4.46	4.07	3.84	3.69	3.58	3.50	3.44	3.35	3.28	3.12	2.93
10	4.96	4.10	3.71	3.48	3.33	3.22	3.14	3.07	2.98	2.91	2.74	2.54
12	4.75	3.89	3.49	3.26	3.11	3.00	2.91	2.85	2.75	2.69	2.51	2.30
14	4.60	3.74	3.34	3.11	2.96	2.85	2.76	2.70	2.60	2.53	2.35	2.13
16	4.49	3.63	3.24	3.01	2.85	2.74	2.66	2.59	2.49	2.42	2.24	2.01
18	4.41	3.55	3.16	2.93	2.77	2.66	2.58	2.51	2.41	2.34	2.15	1.92
20	4.35	3.49	3.10	2.87	2.71	2.60	2.51	2.45	2.35	2.28	2.08	1.84
25	4.24	3.39	2.99	2.76	2.60	2.49	2.40	2.34	2.24	2.16	1.96	1.71
30	4.17	3.32	2.92	2.69	2.53	2.42	2.33	2.27	2.16	2.09	1.89	1.62
40	4.08	3.23	2.84	2.61	2.45	2.34	2.25	2.18	2.08	2.00	1.79	1.51
60	4.00	3.15	2.76	2.53	2.37	2.25	2.17	2.10	1.99	1.92	1.70	1.39
120	3.92	3.07	2.68	2.45	2.29	2.18	2.09	2.02	1.91	1.83	1.61	1.25
∞	3.84	3.00	2.60	2.37	2.21	2.10	2.01	1.94	1.83	1.75	1.52	1.00

data.

To give a specific example; suppose that data of $\hat{\varepsilon} = \varepsilon' - i\varepsilon''$ versus frequency are obtained for a typical biological solution over a limited frequency range. Inspection of the data, or prior knowledge of the substance, may suggest that two possible models could give a satisfactory representation of the data; namely, (i) a single Debye distribution, and (ii) the sum of two single Debye terms. First a suitable fitting program (such as the Marquardt method, § 4.3.2) would be used to obtain the parameters, confidence intervals, and correlation coefficients for model (i). This information would then be carefully studied, and provided that none of the confidence intervals were unexpectedly large or correlation coefficients too great (possibly $|\rho| > 0.8$) then model (i) would be accepted as reasonable and its root mean square error noted. If, however, the average difference between the experimental and calculated values is greater than the expected experimental error, then this could well suggest that model (ii) could be tried and consideration of the magnitudes of the confidence intervals and correlation coefficients would show whether this model is reasonable. However, it is important to notice that even if at this stage model (ii) were to be rejected, the conclusions drawn would not necessarily be that model (ii) was physically unrealistic, but possibly that the data contain insufficient information to resolve two dispersions.

Assuming that both models (i) and (ii) appear reasonable it may be possible to choose between them with an F-test as explained in § 4.4.5. If no significant difference is found then the statistical solution would be to take more data points so that the F-test becomes more sensitive. However, this is often not possible for practical reasons.

In these circumstances decisions may have to be made by utilizing knowledge of previous work. Thus in a practical situation statistics can often only provide a guide, albeit a very useful one, in the correct interpretation of data. However, such inexactness should not be criticized provided that any possible ambiguities in the molecular interpretation are realized and made clear when the results are presented.

To reiterate, when a curve-fitting procedure is being used,

confidence intervals and correlation coefficients *must* be calculated. Any attempt to quote parameter values without this knowledge could yield molecular interpretations which at best are misleading. However, even when a full statistical analysis is performed, care is still needed if the results are to be correctly interpreted.

4.6. Some functions encountered in the analysis of dielectric data

For the calculation of ε' and ε'' from coaxial line results the functions to be fitted are eqns (3.50) and (3.52), which have been discussed in § 3.3.6. Notice, however, that although the equations contain complex algebra the final result is expressed as a modulus and thus the fitting procedure is strictly real.

The automated coaxial line method enables a weighted least-squares curve fit to be used. With such a system each value of voltage can be recorded several times. This enables a variance σ_i^2 to be estimated at each experimental point so that weighting factors of $w_i = 1/\sigma_i^2$ can be calculated (see § 4.3.2(a)).

For the analysis of the resultant dielectric dispersion data, the function to be fitted is usually complex and the form of this function often has to be determined from the experimental data. In § 4.2 we have already mentioned graphical methods which are useful when choosing the correct model. If a single relaxation time is suspected then the appropriate equation is (4.10), or in terms of relaxation frequency f_R,

$$\hat{\varepsilon} = \varepsilon_\infty + \frac{\varepsilon_s - \varepsilon_\infty}{1 + if/f_R} . \qquad (4.24)$$

For a distribution of relaxation times the Cole—Cole function (eqn (4.3)) could be fitted. However, although this function was very valuable in the pre-computer days, another possibility is to use an equation which is physically more reasonable. In § 2.3.3 a distribution of relaxation times has been expressed by eqn (2.50) and could be written in general terms by the following convolution integral:

$$\varepsilon(f) = \int_{-\infty}^{\infty} h(f')D(f-f')df'. \qquad (4.25)$$

It may seem reasonable to take $h(f')$ as a Gaussian distribution symmetric on a logarithmic scale, say $h(v')$, so that

$$h(v') = \frac{1}{\sqrt{2\pi}\sigma_G} \exp\left[-\tfrac{1}{2}\left(\frac{v'-v_0}{\sigma_G}\right)^2\right] \qquad (4.26)$$

and

$$D(v-v') = \varepsilon_\infty + \frac{\varepsilon_s - \varepsilon_\infty}{1 + i10^{(v-v')}}, \qquad (4.27)$$

where v indicates a variable or a logarithmic scale, i.e. $f = 10^v$. Now σ_G^2 is the variance of the Gaussian distribution, $f_0 = 10^{v_0}$ is the mean relaxation time of the broadened dispersion, and v' is a variable of the integration. Although eqn (4.25) contains a convolution integral the usual methods for least-squares curve fitting can be used. However, to evaluate the fitted function the computer now has to perform a convolution integral instead of just calculating a simple algebraic function. This is a straightforward problem since the Gaussian tends to zero as $f \to \infty$ or $-\infty$ and in practice it is sufficient to divide the curve into 100 elements covering 6 standard deviations (i.e. $\pm 3\sigma_G$).

At this stage the relative merits of using σ_G rather than α should be considered. Although the convolution approach has more theoretical backing than the Cole–Cole function, the forms to be taken for $h(v')$ and $D(v-v')$ are by no means clear. In particular, the problems in deducing the forms of the functions h and D for a liquid are insoluble in practice, except in certain restricted cases. Hence normally eqns (4.26) and (4.27) are no more than first approximations.

These problems have also been considered in § 2.3.4. Although it could be considered that the convolution approach is a step in the right direction there is the practical disadvantage that α has been used in most of the previous literature. To overcome this problem Jordan (1971) considered numerically the relation between α and σ_G and from his results Fig. 4.4 has been prepared. It is important to notice that α is not linearly related to σ_G. For any given set of data the relative values of α and σ_G must depend upon the range of the data and the magnitude of the experimental error, so that Fig. 4.4 is not exact.

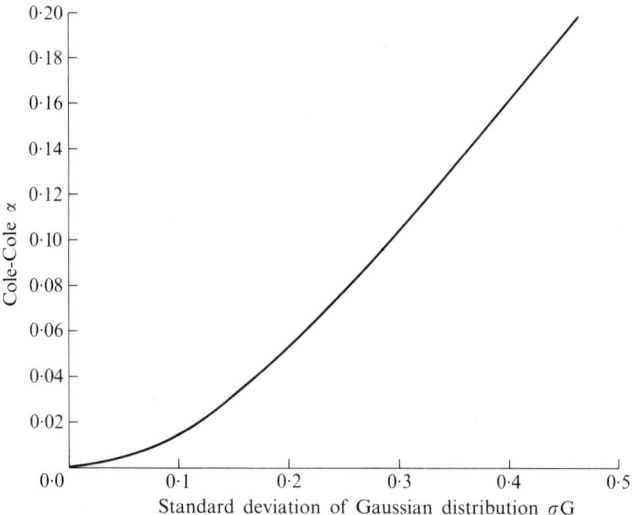

FIG. 4.4. Graph of α against σ_G (based on data from Jordan 1971)

Nevertheless the accuracy should be sufficient for all practical purposes.

When two single dispersions are suspect, then the equation to be fitted is

$$\hat{\varepsilon} = \varepsilon_\infty + \frac{\Delta_1}{1 + i\omega\tau_1} + \frac{\Delta_2}{1 + i\omega\tau_2}, \qquad (4.28)$$

where τ_1 and τ_2 are the relaxation times and Δ_1 and Δ_2 are the increments of the two dispersions. An alternative equation to (4.28) sometimes appears in the literature in the form

$$\hat{\varepsilon} = \varepsilon_\infty + (\varepsilon_s - \varepsilon_\infty)\left\{\frac{c_1}{1 + i\omega\tau_1} + \frac{(1-c_1)}{1 + i\omega\tau_2}\right\}, \qquad (4.29)$$

where $c_1 < 1$ (Bergmann, Roberti, and Smyth 1960).

When the two dispersions are well separated they can easily be fitted. However, as they get closer it becomes difficult, on the basis of the experimental data, to distinguish between two dispersions or a distribution of relaxation times. This problem has been considered recently (Sheppard and Grant 1974) and it was shown that for a typical set of data, two dispersions of equal increment could not be resolved if $\tau_1:\tau_2 < 1:3$. Although this limit of resolution will be a function of the data, care is required in the interpretations when τ_1 and τ_2 are close together.

For three dispersions eqns (4.28) and (4.29) become

$$\hat{\epsilon} = \epsilon_\infty + \frac{\Delta_1}{1 + i\omega\tau_1} + \frac{\Delta_2}{1 + i\omega\tau_2} + \frac{\Delta_3}{1 + i\omega\tau_3} \quad (4.30)$$

and

$$\hat{\epsilon} = \epsilon_\infty + (\epsilon_s - \epsilon_\infty)\left\{\frac{c_1}{1 + i\omega\tau_1} + \frac{c_2}{1 + i\omega\tau_2} + \frac{(1-c_1-c_2)}{1 + i\omega\tau_3}\right\}, \quad (4.31)$$

where $(c_1+c_2) < 1$. Clearly eqns (4.30) and (4.31) can be extended to any required number of dispersions. Again great caution must be exercised to ensure that too many dispersions are not being fitted to insufficient data. If several dispersions are visually detectable in the dispersion curve it is fairly safe to attempt to fit them. However, for overlapping dispersions statistics must be used, as explained in the previous sections, to ensure that spurious dispersions are not being fitted.

For all of the above functions the data points should be weighted whenever possible. With coaxial line and waveguide results, ϵ' and ϵ'' are generally obtained from a least-squares fitting program and their confidence intervals are calculated (see § 4.4.2). The variance of a parameter is proportional to the square of its confidence interval so that weighting factors $w_i = 1/\sigma_i^2$ can be calculated as before. For bridge results the errors are less well defined and the variance has to be estimated from prior knowledge of the accuracy of the bridge and the self-consistency of the experimental results.

5
WATER AND SMALL BIOLOGICAL MOLECULES

The biological importance of water has already been referred to in Chapter 1 where it was shown that the water present in a living system can be conveniently divided into two categories, depending upon whether its structural properties are affected by the presence of biological macromolecules. In this event the water is referred to as 'bound'. On the other hand, water molecules more than a few molecular lengths away from the macromolecule constitute the main bulk of the solution and this water is therefore called 'bulk' or 'free' water. Free water may be expected to have similar dielectric properties to those of pure water, but the relaxation time of bound water must be longer than that of pure water because of the stronger forces linking the bound water molecules to their environment, and *to that extent* bound water may be considered as 'ice-like' (§ 5.1.1).

The dielectric properties of pure water have been fully described in other books (see for example, Hasted 1973; Franks 1972; Eisenberg and Kauzmann 1969), and readers should refer to these to obtain detailed information on this subject. In the present chapter we briefly summarize the principal characteristics of the dielectric dispersion of pure water and explain their importance in relation to the understanding of the dielectric behaviour of aqueous biological solutions. A short account is given of the interesting problem of interpreting the dielectric data in terms of the structure of water, and an attempt is made to distinguish between the areas where agreement has been reached and those where controversy still exists.

Although ice has no biological significance we briefly refer to its dielectric properties as a prelude to the discussion on bound water in § 5.4. The chapter concludes with a detailed description of the dielectric behaviour of aqueous solutions of small biological molecules and of how the complex permittivity may be measured and interpreted.

5.1. Dielectric dispersion curve of water

5.1.1. *General features*

The complex permittivity of pure water has been measured by several independent research groups at frequencies up to 35 GHz. There is much less information available between 35 GHz and a few hundred GHz but at frequencies higher than this measurements have been made on water using free space methods. More recently measurement have been made on the dielectric properties of water by van Loon and Finsey (1975) at frequencies greater than 100 GHz but their results do not affect the broad conclusions which we draw in this chapter concerning dielectric dispersion in water. In the same way recent work by ourselves and by Hasted and his co-workers at frequencies in excess of 70 GHz point to some subtleties in the far infra-red behaviour of water but confirm the principal features of the main relaxation region (Szwarnowski, Sheppard, and Grant 1978; Asfar and Hasted 1977). The form of the dispersion (ε') and loss (ε'') curves at 20°C are shown in Fig. 5.1, where it is seen that the permittivity (ε') falls from a static value of about 80 to a little below 20 at 35 GHz. The loss (ε'') curve exhibits a peak at 17 GHz corresponding to a relaxation time of 9.3 ps.

The static permittivity (ε_s) of water was accurately measured by Malmberg and Maryott (1956) over a temperature range 0–100°C. The result of this study can be expressed as follows:

$$\varepsilon_s = 87.74 - 40.008 \times 10^{-2}\theta + 9.398 \times 10^{-4}\theta^2 - 1.410 \times 10^{-6}\theta^3, \tag{5.1}$$

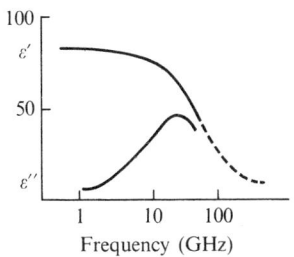

FIG. 5.1. Dielectric dispersion and loss curves for pure water at 20°C

which shows a fairly good approximation to linearity between ε_s and temperature (θ) expressed in degrees Celsius.

The values of ε' and ε'' taken by various workers at frequencies up to 35 GHz have been analysed by Hasted (1973) who showed that the dielectric dispersion of water can be adequately represented by a Cole–Cole distribution equation (eqn (4.3)) with a distribution parameter α lying between 0.01 and 0.02 in the range 0–60°C. Recently, in collaboration with Schwan, we have taken results at 25°C between 100 MHz and 37 GHz and have found $\alpha = 0.014 \pm 0.007$ (Schwan, Sheppard, and Grant 1976). In other work we have suggested that α takes a slightly larger value (between 0.025 and 0.045) near 4°C. Thus, although the Debye equations (eqns (2.44) and (2.45)) are a good first-order approximation of the dielectric dispersion in water, a better description is given by the Cole–Cole equation. The value of ε_∞ found to give the best fit to eqn (4.3) lies in the range 4.0 and 4.5 between 0°C and 60°C. The dotted section of Fig. 5.1 shows ε' falling to a plateau corresponding to $\varepsilon_\infty = 4.3$, but it must be emphasized that this curve is extrapolated from measurements at frequencies below 35 GHz and assumes (wrongly) that there are no other dispersion regions at frequencies higher than this. Hence ε_∞ defined this way is a mathematical parameter chosen to give the best fit of the dielectric data to the Cole–Cole equation; its physical significance is highly sensitive to the model. The problem of ε_∞ is further considered in § 5.1.2.

The relaxation time (τ) of water decreases with increasing temperature in a manner which is approximately exponential. The value of τ at 0°C is 18 ps, falling to 4 ps at 60°C. The equation relating relaxation time with absolute temperature may be written

$$\tau \propto T^{-n} \exp(\Delta H/RT), \qquad (5.2)$$

where n equals 0, ½, and 1 for the Arrhenius, Bauer, and Eyring (eqn (2.55)) equations respectively (Hill et al. 1969). The activation enthalpy (§ 2.3.3) is represented by ΔH. Using relaxation times published by Hasted (1973) we have plotted $\log \tau T^n$ against $1/T$ for $n = 0$, ½, and 1 and have obtained values of ΔH = 18.3, 16.3, and 15.3 kJ/mol respectively at 30°C. In none of the cases was a linear relationship exhibited but the closest approximation to a straight line was shown in the Arrhenius plot (Fig. 5.2). The curvature becomes more pronounced as n increases and the value of ΔH at any given temperature must be obtained from the gradient by drawing a tangent at the appropriate point. Since the uncertainty in ΔH arising from experimental error is only ± 0.6 kJ/mol, the importance of specifying how an activation enthalpy is derived is apparent.

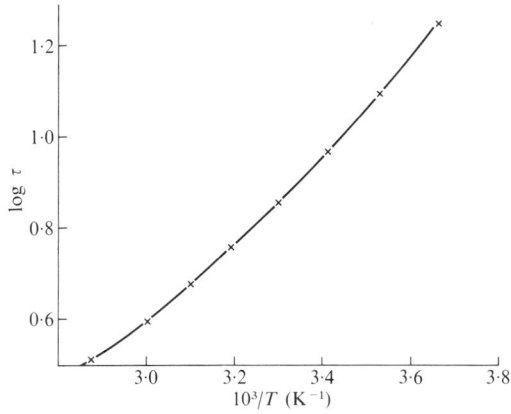

FIG. 5.2. Arrhenius plot for pure water (τ in ps)

Ordinary ice, which should be more correctly referred to as ice I to distinguish it from the other forms produced at high pressures, has a static permittivity of 92 at 0°C and a value of ε_∞ = 3.1. These values are quite close to those of the corresponding parameters of liquid water at the same temperature (ε_s = 87.7, ε_∞ ~ 4.5). The rather surprising similarities in the dielectric parameters of the liquid and solid state may be contrasted with the large differences between the relaxation times. The value of τ for water at 0°C is 18×10^{-12} s and for ice at the same temperature, 2×10^{-5} s. The other important dielectric parameter in which a substantial difference occurs is the activation enthalpy for dielectric relaxation, which is 18.3 kJ/mol for liquid water and 55 kJ/mol for ice (4.4 and 13 kcal/mol respectively). Like liquid water, the dielectric dispersion occurring in ice can be represented to a fair approximation by the Debye equations.

5.1.2. *The meaning of* ε_∞

In the derivation of eqns (2.22) and (2.24) it is assumed that ε_∞ is due to distortion polarization only and therefore that effects due to dipolar rotation are specifically excluded. Since the dispersion region shown in Fig. 5.1 would level out to a value between 4 and 4.5 at frequencies high relative to the relaxation frequency, there must be further dispersion occurring between microwave frequencies and optical frequencies, where the square of the refractive index is only 1.7. The question that has to be decided is whether this fall in permittivity from above 4 to 1.7 is due to dipolar rotation, to intermolecular and intramolecular vibrations, or to a combination both of vibrational and rotational effects. If this fall in permittivity is due entirely to vibrational effects then the true value of ε_∞ is obtained from extrapolating the microwave dispersion data, and lies between 4 and 4.5. Conversely, if vibrational effects are negligible, then $\varepsilon_\infty = 1.7$ and the entire fall in permittivity from above 80 to 1.7 is due to dipolar rotation. More data will be required in the far infrared to settle the question one way or the other. For the interpretation of the dielectric behaviour of biological molecules in aqueous solution the value of ε_∞ and its physical significance are unimportant because the relaxation of the solute molecules takes place at frequencies far below those where $\varepsilon' \sim \varepsilon_\infty$. In other words the value of permittivity ε' in the dispersion region of a biological molecule (other than water itself) is such that $\varepsilon' \gg \varepsilon_\infty$. Hence when dielectric data taken on biological solutions are being analysed and a value of ε_∞ is required, it is adequate to assume that ε_∞ for the water component is 4.3 with no variation in temperature between 0°C and 50°C, and we shall adopt this practice when interpreting the results on amino acids and proteins in the subsequent sections of this book. The magnitude of ε_∞ is, however, very important when interpreting the dielectric data on pure water, and this will be referred to in § 5.2.3.

5.2. Structural interpretation of the static permittivity (ε_s)

5.2.1. General considerations

As explained in § 2.2.3 the static permittivity of a liquid can be expressed in terms of various structural parameters using an equation of the form (2.22). For water $\varepsilon_s \gg \varepsilon_\infty$, from which it readily follows that eqn (2.22) may be re-written as

$$\varepsilon_s = \varepsilon_\infty + \frac{N}{2\varepsilon_0 V} \frac{g\mu^2}{kT}. \qquad (5.3)$$

To use eqn (5.3) in this form it must be assumed that g is the average structural correlation parameter and that μ is the mean molecular dipole moment in the liquid. If the water molecules are all in the same state of bonding with identical average molecular environments, the assumption that g and μ do not vary appreciably throughout the liquid is meaningful, but if wide variations are likely to occur then any mean values adopted for μ and g have limited physical significance. For water the situation is complicated by ignorance of its structure but, despite this, simplifications can be made which allow g and μ to be calculated. Before discussing these we shall briefly describe the configuration and charge distribution pertaining to the isolated water molecule.

In the vapour state the water molecule has an angle of about 105° between the O–H bonds. The positive charge is located in the region of the two protons and the negative charge on two lone pair orbitals on the side of the oxygen atom remote from the protons. The lobes of negative charge are directed above and below the plane containing the oxygen nucleus and the two hydrogen nuclei. A rough idea of the charge distribution can be gained by considering the molecule as a trigonal pyramid with the positive and negative charges located at the apices, as shown in Fig. 5.3. The LOL plane is perpendicular to the HOH plane and the angles LOL and HOH are both about 105°. The dipole moment of the water molecule in the vapour

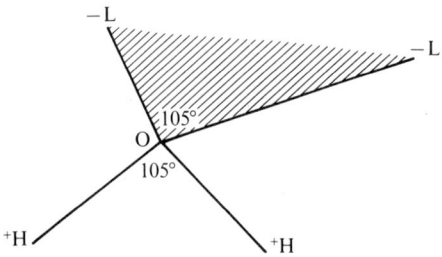

FIG. 5.3. H_2O molecule as a trigonal pyramid. Shaded plane is perpendicular to paper

state has been measured by a variety of techniques to give a value μ_g = 1.84 D.

In liquid water the molecules are linked by the formation of hydrogen bonds (Pauling 1960; Pimental and McClellan 1960). The simplest case is to consider all the water molecules to be four-bonded with each oxygen atom at the centre of a tetrahedron containing four other oxygen atoms, one at each corner of the tetrahedron. The oxygen atoms are separated by a distance of 0.276 nm and the protons are situated such that an O—H···O hydrogen bond is formed between each pair of water molecules. The O—H bond length is about 0.096 nm, which means that the hydrogen atom is roughly twice as far from one oxygen atom as it is from the other. Since the angle subtended at the centre of a regular tetrahedron by one of its sides is 109°, it is seen that the bond angles shown in the simple picture of the isolated water molecule (Fig. 5.3) have to undergo little change when the molecules become part of a network in liquid water. The above description of the form of the isolated water molecule is oversimplified and the assumption that all the molecules in liquid water are four hydrogen bonds is incorrect, but this simple picture provides a good starting point for the interpretation of the static permittivity in terms of molecular parameters from eqn (5.3).

If the interaction of the central molecule with the four neighbouring molecules only is considered (Fig. 5.3) it is easy to show, following Kirkwood (1939), that $g = 1 + 4 \cos^2 54°30' =$

2.13. In this calculation the HOH angle of 105° has been replaced by the tetrahedron angle of 109°. It is interesting to compare this value of g with that obtained from a more sophisticated approach. For a four-bonded molecule in ice, Coulson and Eisenberg (1966) have calculated the molecular dipole moment μ = 2.6 D on the basis of a multipole model and by taking into account interactions of the molecule in question with 85 surrounding molecules. Assuming ε_s = 92 and ε_∞ = 3.1 for ice at 0°C and substituting into eqn (5.3), the value g = 2.6 is obtained. Considering that Kirkwood's model involved interactions with only four neighbouring molecules the general nature of his approach is seen to be justified. In § 5.2.2 we describe how the relationship between ε_s, g, and μ may be shown using an approach based on more modern ideas of water structure.

5.2.2. *Interpretation of the static permittivity of water in terms of the dipole moment of the molecule in the liquid*

Modern theories of water structure date from Bernal and Fowler (1933) who pointed out on the basis of the X-ray diffraction data that liquid water, like ice, contains molecules which are tetrahedrally bonded and therefore that the hydrogen bond plays a predominant role in the structure of liquid water. This fact is now accepted and is consistent with all experimental data on water, whether thermodynamic, mechanical, electrical, magnetic, or spectroscopic. The difference between water and ice could therefore be explained on the basis that considerable bond bending takes place in water (see for example, Pople 1951; Bernal 1964), or that water is characterized by a significant proportion of broken bonds and by the presence of some molecules which are zero bonded (Haggis, Hasted, and Buchanan 1952; Danford and Levy 1962). These two categories of model may be referred to as 'uniformist' and 'mixture' models respectively (Franks 1972). The difficulties involved in interpreting the experimental data on water are underlined by the fact that in one authoritative textbook on the subject, Frank (1972) is arguing in favour of the principle of the mixture model and in another Eisenberg and Kauzmann (1969) are sympathetically disposed towards the distorted hydrogen-bond type of model. However, as far as the dielectric properties are concerned, the fact that ε_s for water at 0°C is only 6 per cent less than that for ice indicates that only a small percentage of bonds are broken or, alternatively, that the degree of bond distortion is small.

Eqn (5.3) must therefore be modified to make provision for the presence of molecules other than those which are rigidly bonded to four neighbours. One way of doing this has been suggested by Haggis *et al.* (1952). In this approach water is considered as a statistical assembly of molecules forming 0, 1, 2, 3, or 4 bonds with the surrounding neighbours. A similar model was proposed by Nemethy and Scheraga (1964) to account for the thermodynamic properties of water.

In the Haggis theory it is assumed that the probability of a hydrogen bond breaking or forming is the same for all bonds and that a certain percentage of bonds is broken at any given temperature. Given this assumption, Haggis *et al.* (1952) showed that

$$\varepsilon_s = \varepsilon_\infty + \frac{N}{2\varepsilon_0 V} \sum_{i=0}^{i=4} \frac{n_i}{100} \frac{g_i \mu_i^2}{kT} . \qquad (5.4)$$

The values of n_4, n_3, n_2, n_1, and n_0, corresponding to the numbers of molecules in the different bond-state populations at any given temperature can, in principle, be calculated in terms of the percentage (p) of broken bonds occurring in water at that temperature. The Kirkwood correlation parameter g_4 for a four-bonded molecule can be obtained by considering the interaction between a central water molecule and its neighbours; from the definition $g_0 = 1$, the values of g_3, g_2, and g_1 can be interpolated. In a similar manner μ_3, μ_2, and μ_1 may be interpolated between the value of μ_4 calculated from a structural model ($\mu_4 = 2.6$ D — § 5.2.1) and the value of μ_0, which may be taken as approximately equal to the dipole moment of the molecule in the gas ($\mu_g = 1.84$ D). It is difficult to calculate accurately the asbolute values of n_i at any given temperature because of ignorance of the percentage (p) of bonds broken, but it is easier to obtain reliable information on the temperature variation of this parameter by the use of thermodynamic data

(Haggis et al. 1952). Therefore the best way of testing eqn
(5.4) is to compare the predicted temperature variation of ε_s
with that observed experimentally. According to the experimental data of Malmberg and Maryott (1956) the value of ε_s at
100°C is 63.5 per cent of its value at 0°C. The corresponding
figure predicted by eqn (5.4) (fitting the values of n_i to give
agreement in ε_s for liquid water at 0°C) is 60.0 per cent,
which shows very satisfactory accordance between theory and
experiment for a liquid of structure as complicated as water.
Agreement at temperatures between 0 and 100°C is also good.
Additionally, support for eqn (5.4) comes from the agreement
which can be obtained between theory and experiment at 4°C for
the *absolute* value of ε_s if values of n_i calculated from
p = 17 per cent are used (Grant and Sheppard 1974a). This value
of p follows from the latent heat data, assuming that one
quarter of the heat of sublimation of ice is due to van der
Waals forces (Pauling 1960).

The bond-bending theory of Pople (1951) predicts a temperature variation in ε_s which is also in excellent agreement with
experiment, and although the absolute values of ε_s calculated
from his theory are all too low, agreement can be reached
simply by choosing a higher value for the molecular dipole
moment. Evidently we may conclude that a theory involving
either bent bonds or a small proportion of broken bonds can be
used to explain satisfactorily the relationship between static
permittivity and structural parameters for liquid water, at
least over the temperature range of interest for biological
solutions. Thus the static permittivity may be accounted for
in terms of the rotation in an electric field of single molecules with a dipole moment of the correct magnitude as calculated from structural considerations, and with a degree of correlation which can be also justified on structural grounds. In
view of the relatively low viscosity of liquid water, it might
be argued that at least some bonds must be broken and to that
extent we would favour the bond-breaking model. More important
than trying to decide between these two possibilities is that
it is not necessary to postulate the existence of aggregates
or clusters to account for the static permittivity of liquid
water. Moreover, in none of the published theories involving

these concepts is there any prediction either of the absolute value of ε_s or of its temperature variation in terms of structural parameters. The weakness of the Pople theory is that 'it ... is clearly too simple to give a good representation of the forces between water molecules' (Eisenberg and Kauzmann 1969). The principal shortcoming of Haggis's bond-breaking theory is the assumption that the probability of a hydrogen bond breaking or re-forming is the same for all bonds. It is much more likely that the strengths of the various bonds which one molecule makes with its neighbours are inter-dependent. However, for water at temperatures of biological interest it would appear likely that the majority of the molecules are four-bonded on the grounds that the static permittivity of liquid water at 0°C is only a few per cent less than that of ice, and even at the body temperature (37°C) the value of ε_s for liquid water is still 80 per cent that of ice. Therefore the proportion of broken bonds is small at these temperatures and the effect of the assumption of uniform bond strengths on the calculated values of n_0, n_1, n_2, n_3, and n_4 may not be too serious. Some of the above arguments have been modified and refined in other publications (Del Bene and Pople 1970) but there is still a serious lack of agreement on the interpretation of the dielectric properties in terms of structure and it is fortunate that these disagreements do not affect the efficiency of the dielectric method of studying biological molecules in aqueous solution.

5.2.3. *Other ways of interpreting the static permittivity of water*

In § 5.2.2 an attempt was made to express ε_s in terms of the dipole moment of the molecule in liquid water. An alternative approach is to use eqn (2.24), which for liquid water ($\varepsilon_s \gg \varepsilon_\infty$) reduces to

$$\varepsilon_s = \varepsilon_\infty + \left(\frac{\varepsilon_\infty + 2}{3}\right)^2 \frac{N}{2\varepsilon_0 V} \frac{g\mu_g^2}{kT}. \tag{5.4a}$$

This equation is equivalent to eqn (5.3) except that the dipole moment of the molecule in the liquid (μ) has been replaced by the gas dipole moment (μ_g) using the expression $\mu_g(\varepsilon_\infty + 2) = 3\mu$, which is eqn (2.23). Eqn (5.4a) has the disadvantage that the relationship between ε_s, μ_g, and g is highly dependent on ε_∞

and also that g is designated merely as the average value of the correlation parameter throughout the liquid without any attempt being made to discriminate between different categories of water molecule as in eqn (5.4). Against this, eqn (5.4a) is superior in its use of μ_g (which is an accurately known parameter) rather than μ, which has to be calculated from the liquid structure. If it is assumed, following Hill (1963), that the value of ε_∞ = 4.3 obtained by extrapolating the principal dispersion is all due to distortion polarization, then the magnitude of g given by substituting the values of the appropriate parameters into eqn (5.4a) is unity at 25°C. This conclusion that g = 1 implies that there is no correlation at all between the molecules and that the effect of hydrogen bonding in water is merely to increase the distortion polarization over and above what might be anticipated for a non-associated liquid. In view of the fact that other non-spectroscopic properties of liquid water require a high degree of correlation for their interpretation, the proposition that g = 1 seems rather unlikely.

A different approach was taken by Zafar, Hasted, and Chamberlain (1973) who proposed that a second relaxation process occurs in the far infrared, with a fall in ε' from 4.3 to 1.8 and a relaxation frequency of 3000 GHz at 25°C. In this case the appropriate value of ε_∞ is now 1.8, which when substituted into eqn (5.4a) gives g = 2.8. This compares well with the value g_4 = 2.66 used by Haggis *et al.* (1952) and with the calculated value of g = 2.60 due to Pople (1951). There are difficulties in assuming that a relaxation process accounts for the value of ε' falling from 4.3 to 1.8 in that the magnitude of ε_∞ for ice is as high as 3.2 and that this has been attributed to intermolecular modes of vibration (Franks 1972). Recent work by Afsar and Hasted (1977) has cast doubt on the relaxation interpretation and it is probably too early to draw any firm conclusions about the dielectric behaviour of water in the far infrared until many more data points are available.

A very promising development which has occurred within the past decade is the use of a computer to simulate the motions of molecules in a liquid. In such a study, Rahman and Stillinger (1971) investigated the behaviour of 216 water molecules assuming a potential function equal to the sum of the potentials of pairs of molecules. They obtained a mean value for the Kirkwood correlation parameter of $g = 2.72$ at 34.3°C, which is in good agreement with the values mentioned in this section and, bearing in mind the limiting nature of the assumed form of the potential function, augurs well for this method of approach to molecular dynamics.

After this short account of the static dielectric behaviour of water and its interpretation, we shall now consider the frequency-dependent properties before finally summarizing the essentials of the complex permittivity and its relation to molecular structure.

5.3. Interpretation of dielectric relaxation in water in terms of molecular motion

The facts to be explained and interpreted are the value of the dielectric relaxation time of water ($\tau \sim 8$ ps at 25°C), its approximate exponential variation with temperature, the associated activation enthalpy of 18 kJ/mol, and the narrow spread of relaxation times. We shall assume (Chapter 2) that the macroscopic dielectric relaxation time can be identified with the molecular correlation time and the symbol τ will be used to stand for either.

The size of the relaxation time suggests that it is related to the movement of individual water molecules rather than to the motion of aggregates or complete regions of the liquid. This proposal can be reinforced by comparing the dielectric data with those obtained from viscosity and diffusion studies of water. Both the diffusion coefficient D and the viscosity η vary exponentially with temperature according to an equation of the form of (5.2), and both processes have an activation enthalpy equal in value to that obtained from the dielectric relaxation time, i.e. $\Delta H = 18$ kJ/mol. This is strong evidence that all three phenomena are determined by the same molecular mechanism. Evaluation of $(\tau D)^{\frac{1}{2}}$, the mean length of a diffusive jump, gives a value of this parameter of 0.37 nm, i.e. of the order of the distance between two molecules (Wang 1965). A linear relationship between η and τT is shown to exist (Fig.

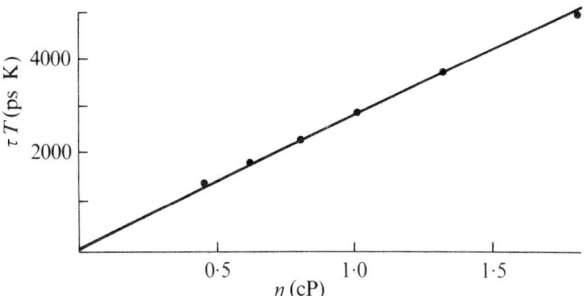

FIG. 5.4. Relationship between viscosity (η), relaxation time (τ), and absolute temperature (T) for pure water

5.4) and this might suggest that the Debye equation (Debye 1929)

$$\tau = 4pa^3\eta/kT \qquad (5.5)$$

holds for liquid water. This equation was deduced from the model that dielectric relaxation of a molecule can be regarded as similar to a sphere rotating in a viscous fluid. In that case the slope of the graph equals $4\pi a^3/k$, where a is the radius of the molecule and k is the Boltzmann constant. The value of a obtained from Fig. 5.4 is 0.14 nm, which is close to half the separation between two molecules in liquid water. It has been shown previously (Grant 1957a) that the linearity between η and τT can be explained in terms of the bond-breaking model for water (Haggis et al. 1952), which was used in § 5.2.1 to interpret the static permittivity results from water. Hence the linearity shown in Fig. 5.4 does not necessarily mean that the assumptions underlying the Debye equation are a true representation of what happens in water, although the magnitude of the slope does lend further support to the idea that the rotating unit is one molecule rather than a cluster. The computer simulation of Rahman and Stillinger (referred to in the previous section) was used to predict the value of the relaxation time of an individual water molecule. They showed that τ = 6.7 ps at 34.3°C, which is not too dissimilar from the experimental dielectric relaxation time of τ = 5.6 ps.

The evidence from the activation enthalpy also favours the unimolecular rotation interpretation. The value of $\Delta H = 18$ kJ/mol is in the energy range corresponding to the breakage of a single hydrogen bond, which therefore suggests that this is the rate-determining step in dielectric relaxation for water. If it were necessary for a whole cluster or domain of molecules to be broken up before orientation were able to take place then presumably many hydrogen bonds would have to be ruptured and ΔH would be higher than 18 kJ/mol. Thus the 'flickering cluster' concept due to Frank and Wen (1957) (to take just one example) would not appear to be a satisfactory model for explaining dielectric relaxation in water. The small observed distribution of relaxation times ($\alpha < 0.02$) means that there cannot be a small number of different sized aggregates each rotating with its own characteristic correlation time. The fact that water very nearly obeys the Debye dispersion indicates that the molecular mechanism underlying the dielectric relaxation must be nearly uniform throughout the liquid, even though it is known from the data on the O—H stretching modes (Walrafen 1968) that there is variation in the molecular environments.

One possible way of accounting for the small distribution of relaxation times is the proposal that there are two similar types of molecular process responsible for dielectric relaxation and that these produce two close but distinct relaxation times (Grant 1956; Hasted 1973). Because the Debye dispersion equations represent permittivity changes occurring over such a wide frequency range, an addition of two Debye processes would appear experimentally to give the same form of dispersion as a small distribution of relaxation times. With typical experimental error it is not possible to discriminate between a distribution of relaxation times with $\alpha < 0.1$ and a sum of two single processes of equal amplitude with relaxation times separated by a ratio less than 1:3 (Chapter 4). In the context of water, one possible explanation of the small observed departure from Debye behaviour is the suggestion that there is a difference between the hydrogen bond strengths associated with symmetrically 2-bonded and asymmetrically 2-bonded water molecules (Grant and Sheppard 1974a). According to this picture the rate of formation of 1-bonded molecules from 2-bonded molecules is the mechanism underlying dielectric relaxation. In that case the 0- and 1-bonded molecules would rotate separately from the main dispersion and at higher frequencies (Hasted 1973, Grant and Sheppard 1974a). Garg and Smyth (1965) found that the relaxation time of water molecules in benzene is 1 ps at 20°C. There would be some similarity between the forces opposing rotation for water molecules in benzene and those appropriate to 0- and 1-bonded molecules in liquid water. The relaxation time of 1 ps (ten times less than that for the principal water dispersion) would therefore give rise to a small dispersion region located in the right part of the frequency spectrum to account for the difference between the permittivities observed in the microwave and far infrared region.

Further support for the bond-breaking model comes from the observation that the activation enthalpy and heat of sublimation of D_2O are a little higher than the corresponding values for H_2O and that the ratios of the relaxation times at any given temperature (Hasted 1973; Grant and Shack 1969) can be accounted for in terms of these differences. Another instructive comparison is between the relaxation times of water and ice at 0°C. The ratio of these quantities is 1.1×10^6. The mean activation enthalpies of water and ice are 55 and 18.3 kJ/mol, which makes the ratio of $\exp(\Delta H/kT)$ 5.6×10^6, in reasonable agreement with that observed experimentally. The magnitude of the activation enthalpy of ice is such as to suggest that three hydrogen bonds are broken when reorientation takes place. It has, however, been more common to attribute the relaxation in ice to the movement of lattice defects (Bjerrum 1951). More recently Minton (1971) has proposed a defect theory to account for the dielectric relaxation in liquid water and has predicted from structural considerations the approximate Debye dispersion behaviour which water exhibits experimentally. Hasted (1973) has pointed out that it is unlikely that the whole of the water dispersion can be accounted for in terms of defects although they might be responsible for the further dispersion which exists in the far infrared.

In the light of the description of the dielectric behaviour

of pure water given in this chapter, our conclusions may be summarized as follows. The dielectric properties of water can be satisfactorily accounted for in terms of the motion of single molecules with a dipole moment and relaxation time of appropriate size for one molecule. The degree of interaction between molecules is shown by the Kirkwood correlation parameter taking a value between 2.5 and 3 in contrast to the value of unity which it would take in the absence of any correlation between the motion of the molecules. There is no need to postulate the existence of aggregates or clusters with their own distinctive characteristics, and to quote Eisenberg and Kauzmann (1969) in their own conclusions on self-diffusion, viscosity, and dielectric relaxation in water: 'No data seem to be inconsistent with a mechanism that involves jumps of individual water molecules.' Whether aggregation is necessary to account for other physical properties of water is outside the scope of this book; clearly it is felt by certain workers in the field that this type of model is necessary to account for some of the spectroscopic and thermodynamic data. It will probably be a long time before the question of the structure of water is settled to the general satisfaction of everyone involved in its study.

5.4. Dielectric properties of bound water

By the very nature of its definition (§ 1.4) bound water cannot be subjected to a direct investigation of its dielectric behaviour. Any conclusions about the dielectric properties of bound water must be inferred from a study of the appropriate solution or suspension. This limits the amount of precise and *absolute* information which can be obtained about its electrical properties, for the following reasons. First, the quantity of bound water present in most biological solutions and tissues is small compared with the amount of 'free' or 'bulk' water. Therefore any changes in the permittivity of the bound water with frequency, temperature, or any other physical parameter take place against the background of the bulk water. In an investigation of the permittivity of bound water, the well-known experimental problem of measuring the difference between two large quantities is encountered. Thus in Fig. 5.5, which is

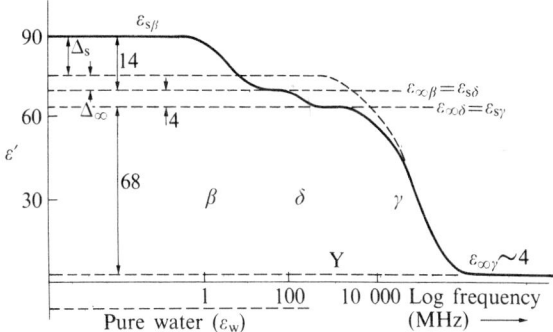

FIG. 5.5. Dispersion curve for 10 per cent solution of myoglobin at 20°C

for a 10 per cent solution of myoglobin, it is seen that the increment of the δ-dispersion is less than 4 compared with the increments due to the protein and free water of 14 and 68 respectively. The second limitation occurs in the interpretation of the dielectric data taken on the biological solution. Some model for the hydrated macromolecule must be assumed on which a suitable dielectric mixture theory for interpreting the permittivity of the solution in terms of the permittivity of its constituents can be based. A further source of error arises in the uncertainty surrounding the values of the dielectric constant and loss of the macromolecule and of the free water at the particular frequency and temperature of measurement, which are required in order that the dielectric parameters of the bound water can be calculated from those of the mixture.

In this chapter we discuss critically the nature of these objections and, in spite of them, draw some conclusions about how permittivity measurements on a biological solution can be used to obtain useful information about the bound water. Some numerical examples explaining the calculation of the hydration (W) are given in §§ 5.4.4 and 6.3.1. In this book our principal concern is to describe the dielectric behaviour of aqueous *solutions* of biological molecules rather than crystals or powders on the grounds that the former state is a closer approximation to the environment of the molecules when *in vivo*. However, because it is easier both experimentally and theoretically to obtain information about hydration from measurements on wet powders, some results obtained from this kind of investigation will be discussed in § 5.4.4. It must be emphasized that the properties of the bound water associated with wet powders are not necessarily the same as those of the bound water in solution and therefore any attempts at comparison must be considered in the light of the model adopted.

5.4.1. *The δ-dispersion and its interpretation*

The dispersion curve of a solution of haemoglobin is shown in Fig. 1.1. Although the nature of the dispersion curve has been previously referred to in Chapter 1 it will be useful to emphasize the details again. The dispersion region between 0.1 MHz and 10 MHz is due to the relaxation of the haemoglobin molecules, and the microwave dispersion occurring at frequencies greater than 1000 MHz corresponds to the bulk water and has dispersion parameters approximately equal to those of pure water (§ 5.1). Between 10 and 1000 MHz a continuous fall in permittivity occurs which can be conveniently represented by a small dispersion. The three dispersions are designated β, δ, and γ respectively and form one continuous curve; thus $\varepsilon_{\infty\beta} = \varepsilon_{s\delta}$ and $\varepsilon_{\infty\delta} = \varepsilon_{s\gamma}$. The existence of the δ-dispersion has been confirmed in egg albumin, bovine serum albumin (BSA), and myoglobin (Fig. 5.5), which with haemoglobin are the only four protein molecules to have been investigated as a continuous function of frequency between 10 and 1000 MHz. Recent work on LDL lipoprotein (Essex 1976; Essex, Grant, Sheppard, South, Symonds, Mills, and Slack 1977) also shows a fall in permittivity with frequency over this frequency range.

It has been proposed by Schwan (1957, 1965) for haemoglobin, by Grant (1962, 1965, 1966) for egg albumin and BSA, and by Grant, Mitton, South, and Sheppard (1974) for myoglobin that the δ-dispersion is due to the relaxation of the bound water present in the solution. Eqns (5.3) and (5.4) show that the macroscopic dielectric parameters of water are dependent upon the environment of a water molecule as well as upon the properties of the molecule itself. Examples of the effect of local environment can be seen in the large difference between the relaxation time of ice and liquid water at 0°C and in the smaller differences between the static permittivity at the same temperature. In an aqueous solution of macromolecules the water adjacent to the solute molecule will clearly be influenced by the presence of that molecule and therefore would be expected to have a correlation time and activation enthalpy different from those of the bulk water. Thus the dispersion of the bound water would be expected to occur over a different frequency region from that of free water and therefore should, in principle, be identifiable from a dielectric dispersion measurement made on the solution. The static permittivity of bound water would also be expected to be different from that of free water, but in view of the similarity between the static permittivities of water and ice, it would not be unreasonable to expect the value of ε_s for bound water to lie in the region 80–100. For ice the infinite-frequency permittivity $\varepsilon_\infty = 3.2$, and for water ε_∞ is around 4–4.5, so it would be anticipated that the value of this parameter for bound water lies in this region. Therefore if it is possible to account for the magnitude of the δ-dispersion in terms of a quantity of bound water in accordance with expectation, and with values of ε_s and ε_∞ equal to the above, it is reasonable to conclude that bound water relaxation is the mechanism underlying the δ-dispersion, especially if there are no obvious alternatives. Thus for the δ-dispersion of haemoglobin, Schwan (1965) shows that the permittivity data are consistent with a value of hydration (W) of 0.3 gram bound water per gram haemoglobin, a value of ε_s for bound water of around 80, and a relaxation frequency f_δ about 300 MHz. For BSA, Grant (1966) obtained $W = 0.2$, $f_\delta \sim 100$ MHz, and an activation enthalpy for the δ-dispersion of about 42 kJ/mol (10 kcal/mol). The studies referred to above on myoglobin and egg albumin were also consistent with f_δ taking a value of a few hundred MHz. In these investigations it was assumed that the hydrated protein molecule consists of a protein core surrounded by a homogeneous shell of bound water of uniform thickness. Thus the parameters W, ε_s, and ε_∞ are the average values for this layer. Also the definition of 'bound' water in this case means that the water molecules are unable to turn in an electric field at frequencies near 1 GHz and therefore contribute to the permittivity of the solution only through their atomic and electronic polarization at this frequency. The value $\varepsilon_s \sim 80$ is appropriate to about 10 MHz, which therefore implies that bound water exhibits its static permittivity at this frequency and below.

Apart from bound water, two other possible mechanisms can give rise to dispersion in the 10–100 MHz region. These are rotation of the polar side chains and proton fluctuation. Pennock and Schwan (1969) measured the permittivity and conductivity of haemoglobin at five concentrations over a frequency range 1–1200 MHz. On the basis of their analysis the dispersion occurring between the protein and water dispersions was attributed to *two* mechanisms: relaxation of protein side chains below 100 MHz and relaxation of bound water above 100 MHz. The dielectric parameters for the bound water consistent with that portion of the dispersion occurring above 100 MHz were found to be $\varepsilon_s \sim 80$, $f_R = 500$ MHz, and $\Delta H \sim 29$ kJ/mol. These are in satisfactory agreement with the values of the parameters of bound water reported above for other proteins. Thus the contribution from the rotation of the protein side chains can be separated from the contribution due to the bound water. This has been recently confirmed for BSA by Essex *et al.* (1977a).

The phenomenon of proton fluctuation has been considered by South and Grant (1973) who show that two categories of dispersion may be considered. If the protons associated with the macromolecule have a fluctuation time τ_f considerably greater than the rotational correlation time τ_i of the macromolecule, then the two dispersion regions overlap and cannot be separately identified from experimental measurement. Conversely, if $\tau_i \gg \tau_f$, two separate dispersions would occur which could be of the form of the β- and δ-dispersion regions shown in Figs. 1.1 or 5.5, the β-region corresponding to the rotation of the macromolecule and the δ-region to proton fluctuation. For myoglobin, better agreement is obtained between the structural dipole moment and that calculated from the dispersion curve if the calculation is carried out according to the condition $\tau_f \gg \tau_i$ (South and Grant 1972). Also, from the known information concerning the rate constants of groups such as COOH and NH_3 it would be expected that $\tau_f > \tau_i$ for myoglobin. Hence for this and other small globular proteins it seems unlikely that any dispersion at frequencies higher than those at which the macromolecule itself disperses is due to proton fluctuation; for larger macromolecules the situation could be quite different. Another fact against the interpretation of the δ-dispersion in terms of proton fluctuation is the observation made for bovine serum albumin that the magnitude of the δ-dispersion is not strongly dependent on the pH of the solution (Grant, Keefe, and Takashima 1968).

Thus it may be concluded that for small globular proteins (i.e., molecular weight less than about 70 000) the δ-dispersion is due either in part or entirely to the relaxation of bound water, and that useful information on the dielectric properties of bound water can be obtained from studying the δ-dispersion in aqueous solutions of proteins of this type. A less accurate, but more straightforward, way of obtaining information about protein hydration is to deduce it from the depression of the permittivity of the solution below the value of pure water at a frequency near 1 GHz.

5.4.2. *The dielectric decrement near 1 GHz*

The relaxation frequency of pure water is about 17 GHz at 20°C and its dispersion curve is described to a fair approximation by eqns (2.44) and (2.45). Hence it can be readily calculated that at frequencies near 1 GHz the value of ε' is about 0.3 per cent less than the static permittivity ε_s which is a very accurately known parameter. The difference between ε_s for pure water and ε' for a protein solution at a frequency near y (Fig. 5.5) is defined as the dielectric decrement, and this quantity divided by the concentration of the protein solution is referred to as the specific decrement (δ). At frequencies in the region of 1 GHz the only contribution to the permittivity of the solution from the protein molecule is due to its atomic and electronic polarization (ε_p); therefore the maximum value which could be calculated for δ if it were due solely to the protein molecules would be obtained if ε_p is assigned its minimum possible value (unity). In practice the value of δ calculated assuming $\varepsilon_p = 1$ and an appropriate shape for the protein molecules is always less, and sometimes considerably less, than the value of δ measured experimentally. This discrepancy is due to the fact that the bound water, as well as the protein, has a low permittivity at this frequency. If there are x grams of protein present in 100 ml of aqueous solution, the water content will consist of xW grams of bound water of permittivity about 4 and the residue of free water of permittivity near 80. Therefore the magnitude of δ is due partly to the protein (whose concentration is known) and partly to the bound water whose hydration value (W) must be calculated. The magnitude of δ for most biological solutions measured to date is about $0.1 \text{ m}^3 \text{ kg}^{-1}$.

The value of W calculated from experimentally measured values of the decrement δ is rather sensitive to the assumed values of the permittivity and density of the protein and the bound water; in this respect the method is markedly inferior to the technique of calculating W from the amplitude of the δ-dispersion. This is made clear in the calculations presented for myoglobin in § 5.4.4. Where the value of the decrement is useful, however, is when it is required to *compare* the amounts of water bound to similar molecules. For example, we have found that the decrement at 800 MHz (δ_{800}) measured for solutions of LDL lipoproteins depended on the pathogenicity of the serum from which the lipoproteins were isolated (Grant, Sheppard, Mills, and Slack 1972; Essex *et al.* 1977*a*). The samples originated from three different clinical states but the shape, size, and density of the lipoprotein molecules were the same in all three cases. Hence the differences in the specific decrements at 800 MHz could be directly interpreted as differences in the hydration. In another investigation δ_{800} was measured for myelin fibres in solution (Gent, Grant, and Tucker 1970) and found to be about three times the value of δ_{800} obtained for globular proteins in solution. The myelin suspension is a difficult system to analyse, but the magnitude of δ_{800} indicates a large quantity of immobilized water.

Although the accuracy in W is poor when calculated from δ, the fact that δ cannot be accounted for solely in terms of the protein and the bulk water indicates, at least qualitatively, the presence of bound water in biological solutions. As stressed already in § 1.4 the term 'bound' in this context means that the water molecules are unable to rotate at frequencies near 1 GHz, in contrast to the free water molecules which can turn freely. This idea of obtaining the hydration from the magnitude of δ was introduced by Buchanan *et al.* (1952) in their classic paper which was the first major contribution in this particular field.

5.4.3. *Measurement of bound water from the radio-frequency dispersion*

The third way in which the properties of bound water may be inferred from the dielectric behaviour of a biological solution is through the study of the β-dispersion (Fig. 5.5). The radius of the rotating unit calculated from eqn (5.5) turns out to be significantly greater than the radius of the macromolecule obtained from X-ray diffraction studies. This effect has been observed for haemoglobin (Grant, South, Takashima, and Ichimura 1971), myoglobin (South and Grant 1972), and ribonuclease (Keefe and Grant 1974). For all three proteins a layer of bound water 2—3 molecules wide surrounding the macromolecule is indicated. The details of the calculation for myoglobin appear in the discussion of the β-dispersion (§ 6.3.1). In the earlier work of Oncley (1943) the presence of bound water is clearly indicated from the observations made on the β-dispersion of 16 proteins in solution, but it was not possible to quantify the bound water as accurately as in the more recent work due to the absence of any reliable data on the dimensions of the protein molecule. The constancy of the width of the bound water shell for the three proteins studied means that the amount of it expressed as a weight or volume fraction increases as the size of the protein molecule diminishes.

5.4.4. *Calculation of hydration in solution from measured dielectric parameters: comparison with wet powders*

In order to calculate hydration values, W. from the parameters $\varepsilon_{s\delta}$ and $\varepsilon_{\infty\delta}$ (Fig. 5.5) the model of a spherical protein surrounded by a spherical shell of bound water has been used (Pennock and Schwan 1969; Grant *et al.* 1974). The mixture formula due to Maxwell (1892) and Fricke (1925, 1955) for the complex permittivity of a suspension (eqn 6.26) may then be applied to both the protein-hydration system and to the solute-water system in turn, each at a frequency first below the β-dispersion and then above it; this leads to the four equations:

$$\frac{\varepsilon_{s\rho} - \varepsilon_{sh}}{\varepsilon_{s\rho} + 2\varepsilon_{sh}} = \frac{1}{1 + W'} \frac{\varepsilon_p - \varepsilon_{sh}}{\varepsilon_p + 2\varepsilon_{sh}} \quad (5.6)$$

$$\frac{\varepsilon_{\infty\rho} - \varepsilon_{\infty h}}{\varepsilon_{\infty\rho} + 2\varepsilon_{\infty h}} = \frac{1}{1 + W'} \frac{\varepsilon_p - \varepsilon_{\infty h}}{\varepsilon_p + 2\varepsilon_{\infty h}} \quad (5.7)$$

$$\frac{\varepsilon_{s\delta} - \varepsilon_w}{\varepsilon_{s\delta} + w\varepsilon_w} = \frac{c(1+W')}{\rho_p} \frac{\varepsilon_{s\rho} - \varepsilon_w}{\varepsilon_{s\rho} + 2\varepsilon_w} \quad (5.8)$$

$$\frac{\varepsilon_{\infty\delta} - \varepsilon_w}{\varepsilon_{\infty\delta} + 2\varepsilon_w} = \frac{c(1+W')}{\rho_p} \frac{\varepsilon_{\infty\rho} - \varepsilon_w}{\varepsilon_{\infty\rho} + 2\varepsilon_w} , \quad (5.9)$$

where the suffixes s and ∞ imply values at frequencies below and above the δ-dispersion respectively, ρ refers to values relating to the solute (protein + bound water) considered as an isolated unit, h and p refer to the bound water and protein core respectively and ε_w is the static permittivity of water. ρ_p is the density of the protein in g cm^{-3} and W' is the extent of the protein hydration expressed as a volume fraction. It is therefore related to the more usual weight fraction, W, by

$$W' = \frac{\rho_p}{\rho_h} W ,$$

where ρ_h is the density of the bound water. Values must be assumed for ρ_p and ρ_h in order to use these equations; ρ_h may safely be taken as 1 g cm^{-3} without introducing any significant error, but the value of ρ_p is not so easy to establish. For myoglobin, for instance, the reciprocal of the partial specific volume differs by about 12 per cent from the density calculated from the molecular weight and the apparent volume obtained from the X-ray crystallography method. Fortunately, however, the equations are very insensitive to this parameter when they are solved by the first method described below, and its choice affects the final result to a negligible degree. Values used for ε_{sh} and $\varepsilon_{\infty h}$ are 100 and 5, in keeping with the previous discussion of hydration, water, and ice (§ 5.4.1).

There are two independent methods of solving the equations. The most reliable method is to use eqns (5.8) and (5.9) to find an expression for the magnitude of the δ-dispersion ($\Delta_\delta = \varepsilon_{s\delta} - \varepsilon_{\infty\delta}$) and then to eliminate ε_{sp} and $\varepsilon_{\infty p}$ from eqns (5.6) and (5.7). The algebra involved in doing this can become cumbersome and is greatly assisted by a computer calculation. The alternative simpler but less accurate method is to use only eqns (5.7) and (5.9), i.e. to calculate the hydration from the decrement $\varepsilon_w - \varepsilon_{\infty\delta}$; this method is further simplified in that $\varepsilon_{\infty p} \ll \varepsilon_w$ and hence eqn (5.9) can be well approximated to

$$\frac{\varepsilon_{\infty\delta} - \varepsilon_w}{\varepsilon_{\infty\delta} + 2\varepsilon_w} = -\frac{c(1+W')}{2\rho_p}, \qquad (5.9a)$$

from which W' (and hence W) may be simply found. The expression $(\varepsilon_w - \varepsilon_{\infty\delta})/c$ is equal to the specific decrement (δ) referred to in § 5.4.2.

In the case of myoglobin (concentration 99 kg m^{-3}), these two methods yield mean values for the hydration of 0.27 g/g and 0.02 g/g respectively (Grant *et al.* 1974). The former value has an associated error of ± 15 per cent which compares favourably with that obtained with other methods of measuring hydration. A theoretical figure of 0.29 g/g has been calculated by Fisher (1965) for myoglobin, and Rosen (1963) found a value of 0.22 g/g by measurements on dampened powders (Fig. 5.6). The failure of the decrement method to produce a reasonable result has two causes; it is very sensitive to the value of ρ_p which is used, and also the bound water is responsible for only a small part of the quantity $(\varepsilon_w - \varepsilon_{\infty\delta})$.

If conductivity data are available throughout the δ-dispersion, a value for W can be calculated (Schwan 1965; Pennock and Schwan 1969) from mixture equations for conductivity equivalent to eqns (5.6)–(5.9), which follow from taking the imaginary part of Fricke's mixture equation (eqn (6.26)). However, the conductivity method has an added error in that the

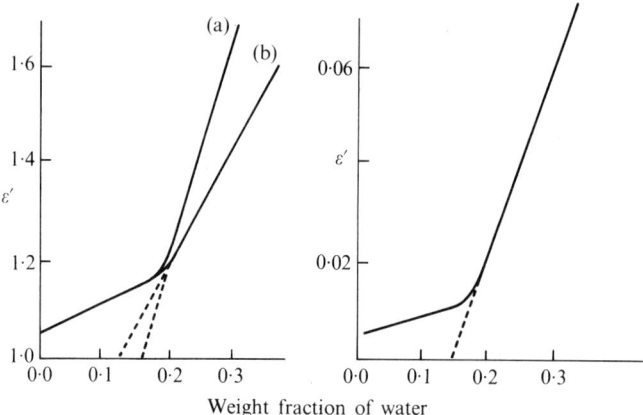

FIG. 5.6. Dielectric properties of packed myoglobin powder as functions of hydration (a) ε' at 10^5 Hz, (b) ε' at 10^6 Hz. ε'' for frequencies $> 2 \times 10^6$ Hz (from Rosen 1963)

conductivity of the protein core may be known much less accurately than the core permittivity (ε_p). Another alternative is to study the variation of conductivity with frequency in the γ-dispersion region. This variation can be used to give information on bound water as was shown by Schwan and Foster (1977) for muscle tissue.

The hydration calculated by eqns (5.6) and (5.7) and by eqns (5.8) and (5.9) refers to the same physical entity. In contrast, the hydration obtained from the calculations based on the β-dispersion refers to the water molecules which are bound to the macromolecule so as to rotate with it as one unit at radio frequencies. This may be considered as 'highly viscous' water, i.e. the differences with free water appear in the mechanical rather than the electrical properties. On the evidence obtained from our studies on haemoglobin, myoglobin, and ribonuclease *this* hydration appears to be 2–3 layers of water molecules for all three proteins and therefore the value of W defined in this way decreases as the size of the protein molecule increases (§ 5.4.3).

Useful information on the hydration of macromolecules has also been gained from dielectric measurements on dampened powders. If the powder is completely dry then the permittivity measured at kHz frequencies is low, generally less than ten, since there are no permanent dipoles that are able to rotate. Rosen (1963) and Takashima and Schwan (1965) found that the permittivity remains low as the protein is dampened until a critical water content is reached. Further dampening results in a marked increase in the permittivity, a typical case being shown in Fig. 5.6. This effect is interpreted by supposing that up to the critical hydration the protein binds each water molecule rigidly, but subsequently water molecules are less tightly bound and are able to rotate in the electric field and thus contribute to the permittivity. Comparison of the bound water levels by this method and that of the dielectric measurement of the solution may be quite good in certain cases; for instance the values of W for myoglobin are 0.25 g/g and 0.30 g/g respectively (Rosen 1963; Grant et al. 1974).

A more detailed analysis has been performed by Brey, Heeb and Ward (1969) on egg albumin and lysozyme. They detected two small discontinuities in the gradients of their capacitance–water content graphs which they attribute to the protein binding water molecules first by two bonds and then, when all such vacancies are filled, by one bond; finally any further water molecules form the second layer of the hydration shell.

Other similar studies on the dependence of permittivity on moisture content have been performed on white fish meal and on protein powders by Kent (1970, 1972), on bone by Freeman (1967) and Marino (1967), on collagen by Tomaselli and Shamos (1973) and on bovine tendon by Lim and Shamos (1971). All these results may be interpreted in terms of the formation of hydration layers around the macromolecules. Comparable studies have been carried out on lysozyme over a wide range of frequencies, and similar conclusions were reached (Harvey and Hoekstra 1972).

5.5. Dielectric behaviour of small biological molecules in solution: Microscopic properties

As mentioned in Chapter 1, it is convenient in any discussion on the dielectric behaviour of biological molecules to divide them into two categories depending upon whether the molecular weight is greater or less than a few hundred. The latter group will be described now and the larger molecules discussed in Chapter 6. Most of the discussion in this section will be concerned with amino acids and peptides, which are an essential group of small molecules present in a living system. A brief description of their structure and a summary of the meaning of the nomenclature is given in Chapter 1 where reference to other small biological molecules of importance will be found.

5.5.1. *The nature of the problem*

The information sought from a dielectric investigation on an aqueous solution concerns the values of the dipole moment and relaxation time of the solute molecule, the form of its dispersion curve, and the effect of any interaction between the molecules and their aqueous environment. Since the solute and solvent molecules are both polar there will be at least two dispersions corresponding to the β- and γ-regions previously referred to for protein solutions (§ 5.4; Figs. 1.1 and 5.5). The principal difference between the dispersion curve of a protein solution and that of an aqueous solution of *small* biological molecules is the considerable overlap between the relaxation regions of the solute and the water which may occur in the latter case. The situation may be exemplified for glycine in aqueous solution. Glycine, the simplest occurring amino acid, has a molecular weight of 75 compared with the value of 18 for water. The moleculer relaxation times of glycine and water at 20°C are about 49 ps and 9.3 ps respectively, i.e. the ratio of the two relaxation times is only five. Therefore the variation of complex permittivity for a 1 M glycine solution is of the form shown in Fig. 5.7.

The real part of $\hat{\epsilon}$ exhibits a fairly smooth variation with frequency, with no visual indication that there are two or more dispersing components present. The curve for ϵ'' shows a small peak near 3 GHz which is roughly the relaxation frequency of the glycine molecule. The maximum in ϵ'' at 17 GHz is due to the water present in the solution. The curves in Fig. 5.7 should be contrasted with those shown in Fig. 5.5 for myoglobin. With the protein solution the β- and γ-regions are so well separated that it is possible to obtain maximum information about the β-dispersion without the necessity for making any measurements at

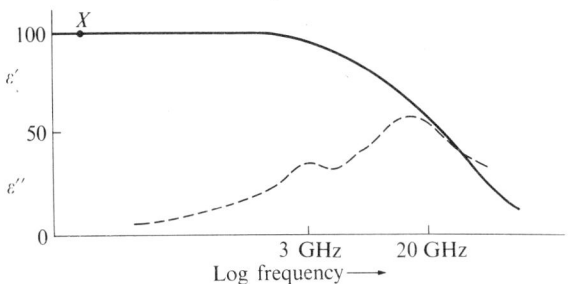

FIG. 5.7. Dielectric dispersion curve of glycine in water at 20°C

all in the water relaxation region. This is the basis of the investigations described in § 6.3. With the amino-acid solution, on the other hand, it is not possible to resolve unambiguously the solute and water dispersions unless highly accurate values of $\varepsilon' - i\varepsilon''$ are available at many frequencies in the region 100 MHz–30 GHz. Thus a comprehensive dielectric investigation on an amino acid solution would ideally require measurements of complex permittivity at a minimum of 100 frequencies in the above range. This would have to be repeated for several different concentrations at a given temperature and the whole process then carried out over a range of temperature. Finally the entire procedure would require repetition as a function of pH in order to identify the mechanism of polarization, e.g. dipolar rotation or proton fluctuation (§ 5.4.1).

The proper interpretation of the results using rigorous methods of analysis is as important as the acquisition of accurate data. It is necessary to analyse the values of $\varepsilon' - i\varepsilon''$ using equations such as (4.29) and (4.31) and employing statistical methods to discriminate between the various apparent interpretations. Any interpretation apparently satisfactory on the grounds of a small root-mean-square error and physically acceptable parameters and errors (i.e. confidence intervals) would then have to be tested to see whether any of the parameters were highly correlated (§ 4.4.3). Although this has already been stressed in Chapter 4, we mention it here again because it is of paramount importance to any system with overlapping dispersion regions.

Unfortunately none of the work published to date on solutions of small biological molecules complies with all of the above requirements. This is hardly surprising since several thousand individual readings of both ε' and ε'' would be required to satisfy the above criteria and several independent sets of apparatus would be necessary to supply the permittivity data. Although these requirements could be met now, the earlier work was carried out over a restricted frequency range and often without adequate computational facilities. If measurements in the time domain prove capable of the same accuracy as those in the frequency domain, then time domain reflectometry (§ 3.6) must have an important future in the field of aqueous solutions of small biological molecules. Although previously published work in this field has its shortcomings, the information and experience gained from it is useful when planning a future investigation in so far as it shows that *some* dielectric parameters can be obtained quite accurately from limited experimental data, while at the same time providing a warning against ambiguous interpretations. Some of these previous investigations will be referred to now as a guide to future research.

5.5.2. *The determination of static permittivity and its use on a simple basis*

A considerable number of experiments have been carried out, some of them more than 30 years ago, on the low-frequency properties of aqueous solutions of amino acids and peptides. The measurements were made at frequencies low compared with the relaxation frequency of the solute molecule and therefore correspond to points on the dispersion curve near X (Fig. 5.7). Tables of results are summarized in Cohn and Edsall (1943) and in Edsall and Wyman (1958). The use of such experimental information is that an approximate value of the dipole moment of the amino acid or peptide can be obtained. Reference to eqn (2.67) shows that the dipole moment is related to the dielectric increment Δ defined by eqns (2.43)–(2.45) and Fig. 2.13. To evaluate Δ accurately it is necessary to know the values of $\varepsilon_{s\beta}$ and $\varepsilon_{\infty\beta}$ (Fig. 5.5). The latter parameter is the value of ε' at which the β-dispersion would level out in the absence of dispersion at higher frequencies. If a reading of ε' is available at one frequency only in the region of X it is not possible to obtain $\varepsilon_{\infty\beta}$ from extrapolation of the β-dispersion, but it can be estimated as follows. If Δ is written as the sum of an increment Δ_s above the permittivity of pure water plus a decrement Δ_∞ below it (Fig. 5.5) then to a first approximation

$$\Delta_\infty \triangleq V(\varepsilon_w - \varepsilon_\infty), \qquad (5.10)$$

where V is the volume fraction (obtained from tables of partial molar volumes) occupied by the solute and ε_∞ and ε_w are the infinite frequency permittivity and static permittivity respectively of pure water. Use of eqn (2.67) in conjunction with eqn (5.10) to calculate molecular dipole moments can be very successful while still preserving the experimental convenience of measuring at one frequency only. This method was adopted by Keefe and Grant (1971) to measure the dipole moment of urea and thiourea in water and ethane diol. Another example of this procedure is shown for triglycine in Table 5.1. The weakness of calculating Δ_∞ from eqn (5.10) is that any contribution from bound water is neglected and that a simple volume proportion mixture theory is assumed. The errors involved in these assumptions are likely to be small and there is the further consideration that $\Delta_s > \Delta_\infty$ and that the dipole moment μ is proportional to the square root of Δ (= $\Delta_s + \Delta_\infty$), which further reduces the uncertainty in μ. Thus for a 1 M solution of glycine in water at 20°C, Δ_s = 22.5 and Δ_∞ = 3.5. If Δ_∞ is in error by 30 per cent, the error in Δ is 4 per cent and in the dipole moment only 2 per cent. In the earlier measurements reported in Cohn and Edsall (1943), the whole of Δ_∞ was ignored and the dipole moment calculated merely from Δ_s. This procedure gives values of dipole moment about 7 per cent too low. In using eqn (2.67) to calculate dipole moments a value must be assumed for the Kirkwood correlation parameters g or, alternatively, an effective dipole moment $g^{\frac{1}{2}}\mu$ may be evaluated. For a protein molecule in solution there are good reasons for taking g equal to unity (South and Grant 1972), but for a smaller molecule this assumption would not be legitimate and it is better to work in terms of the effective dipole moment $g^{\frac{1}{2}}\mu$. Values of this parameter

TABLE 5.1

Effective dipole moments of amino acids and peptides in water at 20°C

	$g^{\frac{1}{2}}_{(D)}\mu$	Method of calculation	Reference
Glycine	19.2	Eqn equivalent to (2.67)	Shack (1972)†
α-alanine	18.6	*	Aaron and Grant (1967a)
β-alanine	20.6	*	Aaron and Grant (1967a)
Proline	20.0	Eqn equivalent to (2.67)	Shepherd and Grant (1968a)
ε-aminocaproic acid	29.8	Eqn equivalent to (2.67)	Shepherd and Grant (1968b)
Diglycine	29.1	*	Aaron and Grant (1967b)
Analylglycine	30.0	*	Aaron and Grant (1967b)
Glycylalanine	30.8	*	Aaron and Grant (1967b)
Triglycine	36.8	Eqn (2.67)	Lawinski, Shepherd, and Grant (1975)
Triglycine	36.4	*	Lawinski, Shepherd, and Grant (1975)

†Based on measurements at 7 kHz (Young and Grant 1968), 190, 250, 700, and 2000 MHz (Young 1967), 450, 600, 760, and 900 MHz (Aaron and Grant 1963), 9.4 GHz (Sandus and Lubitz 1961), and 35 GHz (Shack 1972).

*Eqn (5.10) used to obtain Δ followed by calculation of $g^{\frac{1}{2}}\mu$ from eqn (2.67).

(i) All readings taken near the isoelectric point.
(ii) Solution concentrations: Peptides, 0.25 M; alanine, glycine and ε-aminocaproic acid, 1 M.

for various amino acids are shown in Table 5.1 together with their method of calculation. Further consideration of molecular parameters will be made in § 5.6 after the dispersion properties of solutions have been described.

5.5.3. *Measurement of the dispersion parameters*

Comparatively few papers have appeared describing complex permittivity measurements taken in the dispersion region of amino acids and peptides and over a range of both frequency and temperature. About half of these publications refer to work at three or four isolated frequencies only. The others contain permittivity measurements at a maximum of around 10–15 different frequencies which are largely concentrated in the relaxation region of the solute molecule and therefore provide fairly reliable values of its dielectric parameters, but very little information about other dispersing elements present in the solution. Various relaxation parameters which have been deduced from these types of investigation are shown in Table 5.2, together with a reference to the original publication.

TABLE 5.2.

Relaxation time and activation enthalpy for amino acids and peptides in aqueous solution (20°C)

	Molecular weight	τ (ps)	f_R (GHz)	ΔH (kJ/mol)	Reference
Glycine	75	49	3.23	18.6	Shack (1972) (see Table 5.1)
α-alanine	89	92	1.72	15.9	Aaron and Grant (1967*a*)
β-alanine	89	87	1.84	15.9	Aaron and Grant (1967*a*)
Glycylalanine	146	170	0.94	16.8	Aaron and Grant (1967*b*)
Analylglycine	146	182	0.88	15.9	Aaron and Grant (1967*b*)
Triglycine	189	200+	0.80+	—	Conner and Smyth (1942)
Triglycine	189	258	0.62	18.7	Lawinski, Shepherd, and Grant (1975)
ε-aminocaproic acid	131	284	0.56	18.8	Shack (1972)

(1) Errors - $\tau \pm 10\%$
$\Delta H \pm 2$ kJ/mol (glycine ± 3 kJ/mol)

(2) + Extrapolated from measurements taken at 25°C

(3) Measurements on all solutions near the isoelectric point

(4) Concentrations: peptides, 0.25 M; amino acids, 1 M.

The data on the solutions of 1 M α- and β-alanine were obtained from measurements of ε' and ε'' at frequencies of 100 kHz and 450, 580, 780, and 900 MHz over a temperature range 0–50°C. At each frequency the contribution of the water was calculated from the known dispersion characteristics of pure water; this contribution was then subtracted from the measured values of ε' and ε'' to give the contributions appropriate to the alanine ($\delta\varepsilon'_A$ and $\delta\varepsilon''_A$). The relaxation time (τ) was then calculated from the observed variations of $\delta\varepsilon'_A$ and $\delta\varepsilon''_A$ with frequency, assuming that these variations conform to a single relaxation time. The measurement at 100 kHz provides ε_s directly, but to check the validity of the assumption of a single relaxation time ε_s was evaluated from the data taken at the four other frequencies and was found to agree with the permittivity measured at 100 kHz. As a further test of the validity of the analysis, a plot was made of log τ against the reciprocal of the absolute temperature and a good straight line was obtained (eqn 5.2) with an activation enthalpy of 16 kJ/mol, which is the correct magnitude for a bond breaking mechanism. This alanine study is a good example of reliable information being obtained from a minimum of experimental data, particularly as all five measuring frequencies are well below the relaxation frequency of the alanine molecule; the smaller the part of the dispersion region covered by the experimental point, the more vulnerable the analysis to error. Recent work carried out for valine shows a dielectric behaviour similar to that of alanine (Croom, Shack, Shepherd and Shepherd 1978).

The complex permittivity of a solution of 1 M glycine was measured at certain frequencies in the range 7 kHz–35 GHz by the various research partnerships listed in the footnote to Table 5.1. The data for glycine have been assembled and analysed by Shack (1972). The values of $\varepsilon' - i\varepsilon''$ were fitted to eqn (4.29) and the values of c_1 and τ_1 (the relaxation time of glycine in water) obtained. The dipole moment of the glycine molecule $g^{\frac{1}{2}}\mu$ was calculated from eqn (2.67) by making the substitution $c_1(\varepsilon_s - \varepsilon_\infty) = \Delta$ and the activation enthalpy ΔH obtained from the Arrhenius plot. Values of $g^{\frac{1}{2}}\mu$ and of τ_1 and ΔH are shown in Tables 5.1 and 5.2 respectively.

The triglycine parameters shown in Tables 5.1 and 5.2 were obtained from measurements of $\varepsilon' - i\varepsilon''$ on a M/4 triglycine solution at 24 frequencies between 55–4000 MHz, plus a determination at 1 MHz to give ε_s directly and a point at 35 GHz to provide information at the high-frequency end of the water dispersion (Lawinski 1974; Lawinski, Shepherd, and Grant 1975).

The curves of ε' and ε'' against frequency at 12.5°C are shown in Fig. 5.8. The distribution of experimental points throughout the dispersion region was chosen so that the points were equally spaced on a logarithmic scale between the highest and lowest measuring frequencies (4 GHz and 55 MHz) which are each a factor of ten away from the relaxation frequency (490 MHz). Thus the ratio of the highest to the lowest measuring frequency is about 100, which covers the range of a Debye dispersion (Fig. 2.13). The choice of a frequency range to cover two decades with the relaxation frequency equal to the geometric mean of the two extreme frequencies represents the ideal conditions for obtaining reliable information about a single dispersion in the absence of other dispersions. In this case the dispersion throughout which the experimental points are well distributed is that of the triglycine molecule. However, the lack of data between 4 and 35 GHz means that little information is available on the dispersion of the free water and for the same reason it is not possible to provide reliable evidence for or against an intermediate dispersion. The values of Δ_1, Δ_2, τ_1, and τ_2 were obtained by fitting the data to eqn (4.28). Correlation coefficients were evaluated between the parameters and were shown to be appreciably less than unity in all cases. The values of relaxation time τ_2 were shown to be slightly larger than those for pure water, which is expected for a dilute solution. The analysis was repeated with a Cole—Cole distribution function included for the triglycine dispersion, but the analysis gave $\alpha = 0$.

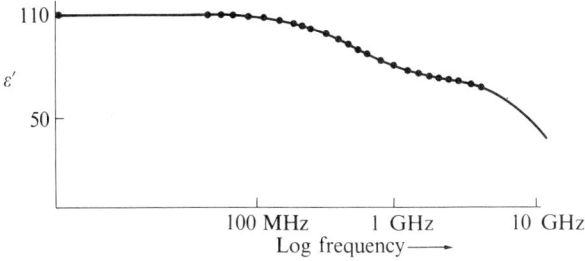

FIG. 5.8. Dielectric dispersion curve of triglycine in water at 12.5°C (from the data of Lawinski *et al.* 1975)

So far we have considered only amino acids and peptides in aqueous solution. Other small molecules of biological interest which have been studied include urea and those occurring in food, none of which are dipolar ions and therefore have a smaller dipole moment (and hence dielectric increment) than amino acids and peptides. Dielectric work on monosaccharides has been carried out using TDS techniques (§ 3.6). These measurements by Tait, Suggett, Franks, Ablett, and Quickenden (1972) are the first published examples of TDS being applied to solutions of this nature. Amongst other substances they measured a 2.8 M aqueous solution of glucose and obtained 22 permittivity values between 300 and 10 000 MHz. The data were interpreted in terms of three dispersion regions with relaxation times of 18.5, 69, and 250 ps at 5°C due to the reorientation of free water, glucose, and bound water molecules respectively. The activation enthalpies were 23 kJ/mol for the glucose molecule and 38 kJ/mol for the water of hydration (compared with $\Delta H \sim 42$ kJ/mol obtained for the hydration water associated with BSA — § 5.4.1). In this study the permittivity data were well distributed throughout the relaxation region of the glucose molecules, whose dispersion parameters are accordingly well defined as a result of the analysis.

Urea is another biological molecule which is not a zwitterion. It has a relaxation frequency of a few GHz. The complex permittivity of solutions of urea in water up to 9 M concentration was measured by Grant, Keefe, and Shack (1972) at 17 kHz (which gives the static permittivity value directly), 9.6 GHz, and 35 GHz. The data were analysed according to eqn (4.29), assuming τ_2 to be the pure water value. The dipole moment of the urea molecule was calculated from eqn (2.67) with Δ replaced by $c_1(\varepsilon_s - \varepsilon_\infty)$.

Nearly all dielectric studies on solutions of biological molecules have been carried out with water as the solvent. This choice has been made principally because water is their natural environment in a living system but also because of the difficulty of finding another liquid in which molecules of biological interest will dissolve. One such exception is proline, which is an amino acid and contains three CH_3 groups, a $\overset{+}{N}H_2$ group and a $CHCOO^-$ group in the form of a ring structure. Its biological importance is briefly referred to in § 1.4. The dielectric dispersions of proline in methanol, formamide, and ethane diol have been measured by Jordan (1971) and in water by Shepherd and Grant (1968a). The values of the dielectric increment (Δ) and relaxation time of proline (τ_p) at 20°C are shown in Table 5.3. All values refer to a 1 M solution. In the same table are shown the relaxation time (τ_s) of the pure solvents. It is noticed that the magnitude of Δ is almost independent of the nature of the solvent, in contrast to the relaxation times which are very dependent on it. Attempt will now be made to interpret this and the other observations made in this section in terms of the structure of the molecules in solution.

TABLE 5.3

Dielectric increment and relaxation time of proline in various solvents

	Molar increment (Δ) (± 2)	$\tau_{proline}$ (ps) (± 15%)	$\tau_{solvent}$ (ps)	Viscosity (cP)
Water	30.0	69	9.3	1
Methanol	27.8	289	56	5.9
Formamide	31.8	334	42	38
Ethane diol	33.2	1200	141	199

5.6. Dielectric behaviour of small biological molecules in solution: Molecular interpretations

5.6.1. *Static permittivity*

In Table 5.1 the values of the effective dipole moment $g^{\frac{1}{2}}\mu$ are shown for several amino acids and peptides. To evaluate the dipole moment the value of g would be required, and to calculate this it would be necessary to have detailed knowledge about the immediate environment of the amino acid or peptide molecule in solution. At present this knowledge is not available and it would be unwise and inaccurate to try to estimate a value for g. For a protein molecule in solution it is safe to assume that g is near unity (South and Grant 1972) and for pure water it has been shown in § 5.2.2 that g is between 2.5 and 3 for a four-bonded molecule. Presumably the value of g for an amino acid or peptide will be shown to lie between these values.

The most effective way of using the information contained in Table 5.1 is to compare the values of $g^{\frac{1}{2}}\mu$ between systems of molecules which are alike. To within experimental error (about ± 5 per cent in $g^{\frac{1}{2}}\mu$ for the molecules listed in Table 5.1) the effective dipole moments of glycine and α-alanine are equal. This reflects the approximate constancy which would be expected in the distance between the amino and carboxyl groups for all α-amino acids (§ 1.3). As this separation increases, the dipole moment also increases, as may be seen from the values of the dipole moment for β-alanine and ε-aminocaproic acid in which there are two and five carbon atoms respectively between the charged end-groups. The peptide dipole moments are also related to the size of the molecule, as would be expected for a zwitterion where the principal factor determining the value of the dipole moment is the separation between the two charged-end groups. The effect of the bond dipole moments is likely to be much smaller; Edsall and Wyman (1958) estimate that each peptide bond contributes about 0.8 D to the total molecular dipole moment.

This increase in dipole moment with chain length has been quantified by Wyman (1936) and by Conner, Clarke, and Smyth (1942) who observed a linear relationship between the dielectric increment Δ_s over that of pure water ($\Delta_s = \varepsilon_s - \varepsilon_w$) and the number of glycyl residues (n) present for all the glycyl peptides up to heptaglycine ($n = 1$ to $n = 7$). This therefore implies a proportionality between the square of the dipole moment and n, which is equivalent to the statement that the mean square distance between the ends of the chain is proportional to the number of atoms (or, for the glycyl peptides, to the number of glycyl residues) present in the chain. This can be interpreted by assuming free rotation about the valence bonds in the chain (Edsall and Wyman 1958). The effective dipole moments of the first three members of the series of glycyl peptides ($n = 1, 2, 3$) are shown in Table 5.1. A plot of $g^{\frac{1}{2}}\mu$ against $n^{\frac{1}{2}}$ for these three molecules gives a satisfactory straight line passing through the origin, thus confirming the results of the earlier work.

The charge separation between the amino and carboxyl groups of the glycine molecule when present in the crystal has been calculated as 0.317 nm (Aaron and Grant 1963), which corresponds to a dipole moment of 15.2 D. Substitution of this into the value of $g^{\frac{1}{2}}\mu$ gives $g = 1.6$, which is intermediate between the values of g appropriate to a molecule in pure water and a protein molecule in solution.

The interpretation presented so far conforms to the idea of single molecules rotating individually in an aqueous environment. This is consistent with the linearity shown between permittivity and concentration. Evidence of this (Fig. 5.9) is given in glycine solution up to concentrations of near 2.0 M (10 per cent by volume). At concentrations above 2.0 M the relationship departs slightly from linearity, possibly owing to molecular association. For urea in water the dielectric increment was found (Grant, Keefe, and Shack 1972) to be almost linear with concentration up to 9 M (volume concentration of urea nearly 50 per cent). Urea is a small polar molecule of molecular weight 60 which forms hydrogen bonds and therefore is expected to exhibit similar dielectric behaviour to that of an amino acid solution.

A further problem in interpreting the dielectric behaviour of aqueous solutions of small polar molecules is whether any

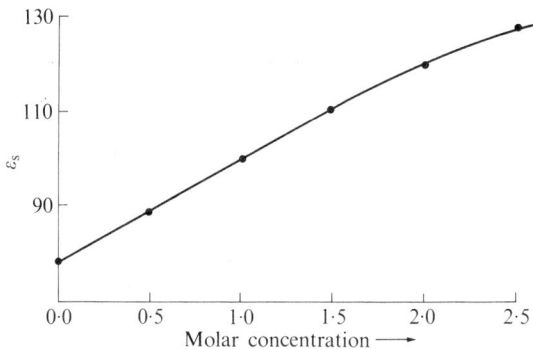

FIG. 5.9. Variation of static permittivity of glycine solution with molar concentration (from Aaron 1966)

particular shape can be meaningfully attributed to the solute molecule. A system of macroscopic spheres in a continuum can be treated mathematically in a rigorous manner and the results applied to biological solutions of practical interest (§ 6.4.4). In contrast, an amino acid solution cannot be considered as a collection of charged particles with well-defined boundaries, distributed throughout a continuum. The effect of considering an amino acid molecule to have an ellipsoidal shape has been investigated by Buckingham (1953), who calculated the dipole moment of β-alanine and obtained a value about 10 per cent less than that in Table 5.1. This fact, taken in conjunction with the values obtained for the dipole moment of various molecules using eqns (2.67) and (5.10), suggests that the model of a rigid dipole rotating in a continuum is a satisfactory one for interpreting the static properties of amino acid solutions and peptide solutions at a molecular level. The independence of the dipole moment of the effect of the surroundings is shown in the first column of Table 5.3, where it is seen that the dielectric increment of proline is largely unaffected by the nature of the solvent.

Evidence against a charge transfer mechanism being responsible for the polarization follows from the absence of free charges in any appreciable extent from the amino acid or peptide solutions referred to in Tables 5.1 and 5.2. All these studies were carried out on solutions near the isoelectric point and therefore the number of protonated and deprotonated states would be too small to give rise to any dispersion effects. Amino acid solutions have a wide isoelectric range owing to the big difference between the pK values of the amino and carboxyl groups (9.7 and 2.3 respectively).

5.6.2. Dielectric relaxation in solutions of small biological molecules

The features immediately apparent from Table 5.2 are the dependence of the relaxation time on the molecular weight and the practically constant activation enthalpy. The molecular weights of analylglycine and glycylalanine are identical and the relaxation times agree to within experimental error. The situation is the same for α- and β-alanine. Conner and Smyth

(1942) measured the relaxation times (τ) of the first five glycyl peptides and found proportionality between τ and the size of the molecule. More recently Croom (1973) showed an approximate linear relationship between relaxation time and molecular weight for 20 biological molecules ranging from water up to haemoglobin (which has a molecular weight of 68 000). Thus it may be concluded, in agreement with the findings in § 5.6.1, that the dielectric behaviour of biological molecules in water can be interpreted in terms of the rotation of molecules individually. The magnitude of the activation enthalpies given in Table 5.2 suggests that the breaking of a hydrogen bond is involved in the dielectric relaxation process.

To test whether the rotation of the molecule in its environment conforms to the Debye equation (eqn (5.5)) a plot was made of τT against η for triglycine (Fig. 5.10). The linear relationship is typical of that obtained for a plot of relaxation time against viscosity for any of the molecules listed in Table 5.2. The triglycine solution was of 0.25 molar concentration, which corresponds to a volume concentration of about 4 per cent. Therefore each triglycine molecule can be considered to rotate in an essentially aqueous environment and the appropriate viscosity is that of pure water. This has been used in Fig. 5.10. If the molecule is spherical, the slope of the graph is $4\pi a^3 k^{-1}$ where a is the radius of the sphere and k is Boltzmann's

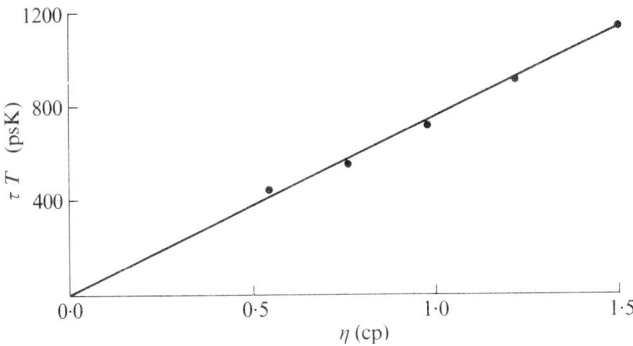

FIG. 5.10. Relationship between viscosity (η), relaxation time (τ), and absolute temperature (T) for triglycine in water (from Lawinski et al. 1975)

constant. If the molecule is non-spherical in shape but can be represented as a spheroid with semi-axes a, b, b, then the slope of Fig. 5.10 contains a shape factor involving a and b (§ 6.3.1). This is another situation which highlights the difficulties of interpreting dielectric data taken on solutions of *small* biological molecules. The particular molecule represented in Fig. 5.10, triglycine, is a short chain of length less than four water molecules and therefore the concept of shape factor, which is so important in interpreting the protein solution data (§ 6.3), has no precise meaning here. However, if $4\pi a^3$ is defined as three times the effective volume of the rotating molecule, the magnitude of a can be evaluated from a plot such as that shown in Fig. 5.10 and compared with other molecular parameters. For triglycine $a = 0.44 \pm 0.04$ nm compared with the extended length of the molecule of 1.0 nm calculated from known bond lengths. The value of $g^{\frac{1}{2}}\mu$ is 36.80 D (Table 5.1) which corresponds to a charge separation of $(0.78\ g^{-\frac{1}{2}})$ nm. Adopting realistic values of g, the minimum distance between the charged ends would be 0.7 nm compared with an effective diameter of 0.88 nm obtained from the graph of relaxation time against viscosity. This observation lends further support to the idea that both phenomena are due to rotation of single molecules in the electric field. As expected the molecule is seen to take up a form in solution which is more compact than the extended chain.

The dependence of the relaxation time on the surrounding environment of a molecule is shown for proline in Table 5.3. The relaxation time of the proline molecule is strongly dependent on the liquid used as the solvent. The ratio of the relaxation time of the proline to that of the solvent lies between about five and eight for the four liquids shown and a rough dependence of relaxation time on viscosity is also observed.

Most of §§ 5.5 and 5.6 have been concerned with the dielectric behaviour of the amino acids and peptides in water and with obtaining information about the structure of these molecules from permittivity measurements. It is also of interest to consider the complementary exercise of investigating the effect of the presence of the biological molecules on the structure

and dielectric behaviour of pure water. In the ideal situation
of permittivity values being available at closely separated
frequencies ranging from the low frequency end of the amino
acid or peptide dispersion up to a few tens of GHz, accurate
parameters appropriate to both the solute and water dispersions come naturally out of the analysis. These conditions
have not yet been realized in practice and it has been traditional to divide the field into the separate studies of the
solute dispersion and the water dispersion, the first category
of study having been the major concern of §§ 5.5 and 5.6 so
far.

To investigate the changes in the water dispersion region
due to a small volume of polar solute molecules many values of
permittivity closely spaced between 2 and 100 GHz would be
required. Unfortunately, the only dielectric investigations
which have been made to date consist of measurements at a
few frequencies only in the above range (Haggis *et al.* 1952;
Delbos and Salefran 1972; Bottreau, Delbos, Marzat, Lacroix,
Salefran, and Dutuit 1973; Bottreau, Delbos, Marzat, and Salefran 1973; Bottreau, Delbos, Marzat, Salefran, and Moreau
1974). In the work of Bottreau *et al.* solutions of glycine,
β-alanine, γ-aminobutyric acid, δ-valine, and ε-aminocaproic
acid were measured at 3.6, 9.5, and 35 GHz at varying concentrations. By combining their results with permittivity data
taken by other workers at lower frequencies they found that the
solute affected the dielectric behaviour of water by requiring
the single dispersion region with relaxation time τ_w to be
replaced by two dispersion regions of relaxation times τ_{w1} and
τ_{w2}, both of which are longer than τ_w. Recent work from the
same laboratory on α- and β-alanine indicates that the dielectric data taken on each solution may be represented by two
components (Salefran, Delbos, Marzat, and Bottreau 1977).

In an analysis of the dielectric data at 20°C on 1 M glycine solutions carried out by Shack (1972), permittivity values
taken between 7 kHz—35 GHz were fitted to eqn (4.29) and the
dipole moment of the glycine and water molecules calculated.
The result for the glycine molecule has been discussed
already in Table 5.1; the value of $g^{\frac{1}{2}}\mu$ for the water molecule
calculated from eqn (2.67) by replacing Δ with $(1-c_1)(\varepsilon_s-\varepsilon_\infty)$

is 3.6 D. In the study of the dielectric properties of solutions of urea referred to in § 5.6.1, measurements were made at 17 kHz, 9.6 GHz, and 35 GHz, and the permittivity results interpreted using eqn (4.29). For the 1 M solution at 20°C the dipole moment and relaxation time of the urea molecule were found to be of the correct magnitude for the dimensions of the molecule and a value $g^{\frac{1}{2}}\mu$ = 3.7 D was obtained for the water molecule in the urea solution. This agrees well with the value of $g^{\frac{1}{2}}\mu$ found for the water molecule in glycine solution and is also consistent with the values of g and μ evaluated from structural considerations for pure water (§ 5.2.2).

6
PROTEINS AND LARGER STRUCTURES

6.1. Introduction

It has been explained in § 1.1 that the dielectric dispersion curve of a typical biological solution can, in general, be divided into four separate dispersion regions. An example of this is shown for muscle tissue in Fig. 1.2, where the α-, β-, and γ-dispersions are illustrated. The origin of the α-dispersion has not been established with certainty but it is thought to be due to a relaxation of counterions or to intracellular structures. The β-dispersion can be divided into two parts. The major part of the fall (tens of thousands to a few hundred) in relative permittivity is due to the inhomogeneous nature of the medium giving rise to large variations with frequency of permittivity and conductivity. This is known as the Maxwell-Wagner effect and is referred to several times in this chapter. The relaxation of macromolecules (e.g. proteins or nucleic acids) accounts for the remainder of the drop in permittivity within the β-dispersion; an example of this is shown in Fig. 5.5 which is for a solution of myoglobin in water.

The properties of the δ-dispersion have already been described in § 5.4. The γ-dispersion exhibited by a protein dispersion is due to the relaxation of the free water but the presence of the protein molecules slightly alters the values of the dispersion parameters from those appropriate to pure water. Although there would be little point now in studying the γ-dispersion in isolation, it is possible to obtain information on solute—solvent interactions from knowledge of the γ-dispersion parameters, as has been shown from earlier work carried out when it was less easy to obtain equipment functioning over a wide frequency range (Buchanan *et al.* 1952; Haggis *et al.* 1952; Grant 1957*b*).

To study the β-dispersion of an aqueous protein solution it is necessary to make measurements in the radio frequency region (a few kHz to around 100 MHz) using bridge methods (§ 3.2). Considerable care needs to be taken over the preparation of samples since they are required to be homogeneous and fairly free from small conducting ions. Any variations in the size of the dissolved protein molecules or the presence of oligomers will have the effect of broadining the β-dispersion, which will lead to misleading deductions. The presence of small ions must be minimized in order to eliminate, or reduce to a tolerable level, the effects of electrode polarization (§ 3.2.2); dialysis against distilled water is not always totally successful in achieving this and alternative methods may be necessary, such as electrodialysis or the use of a Dintzis column. This reduces the salt content of the solution to below the value *in vivo*, usually by a factor of between 10 and 1000.

Of the three dispersions displayed by a protein solution, the β-dispersion has been the most extensively studied since it is amenable to measurement by radio frequency bridges which have been available for years. The instigator of such studies was Oncley, who in the 1930s and 1940s made a series of measurements on such molecules as horse haemoglobin, horse serum-pseudoglobulin, and egg albumin. He interpreted his results on the basis of a theory of the rotation of molecules having a permanent dipole moment, and was thus able to assess the size, shape, and dipole moment of such molecules (Oncley 1943).

Considerable doubt was later thrown upon these deductions by several authors who postulated various mechanisms other than molecular rotation to account for the β-dispersion, as described in § 6.2. However, in recent years several studies have reinforced the model of molecular rotation and it now appears, for the smaller globular proteins at least, that unambiguous conclusions concerning molecular size, shape, and dipole moment may be made from a knowledge of the β-dispersion; if the results of other experiments such as birefringence (see Fredericq and Houssier 1973) or X-ray diffraction are also known then greater weight is given to these deductions.

This chapter continues with a discussion of the various mechanisms which have been postulated to account for the β-dispersion, followed by an account of the procedures involved in deriving molecular information from measured dielectric parameters. It concludes with a brief survey of the use of dielectric measurements in the investigation of larger biological structures.

6.2. The β-dispersion in protein solutions

For most proteins which have been extensively studied the magnitude, Δ, of the β-dispersion is proportional to the concentration of the solution (provided this is not too high) and it is therefore natural to attribute this dispersion to some additional polarization contributed by the protein. There are, in principle, several ways in which this polarization could arise:

(a) *Permanent dipole rotation* (Debye 1929; Onsager 1936; Kirkwood 1932, 1939). This theory supposes that each protein molecule possesses a permanent dipole moment which will experience an orientational force when subject to an electric field. The constant Brownian motion of the molecule will try to randomize this alignment and the magnitude of the additional polarization will depend on the extent of these two effects. Eqn (2.67) applies to this case and since $N/V = Lc/M$ where L is the Avogadro constant, c is the concentration in kg m^{-3} (mg/ml), and M is the molecular weight,

$$\delta = \frac{\Delta}{c} = \frac{L}{2kT\varepsilon_0 M} g\mu_A^2 , \qquad (6.1)$$

where δ is the specific increment.

If an alternating electric field of increasing frequency is applied then the permittivity of the solution will not change appreciably until the frequency approaches the fluctuation rate of the Brownian motion of the molecules; if the frequency is increased beyond this value the molecules will have insufficient time to reorientate between different directions of the field, thus this part of the polarization of the solution disappears and there is a subsequent drop in the permittivity, i.e. a dispersion region. Debye (1929) showed that if the molecule is considered to be a sphere of radius a, and if the solution has a viscosity η, then this dispersion is described by a single relaxation time, given by

$$\tau = \frac{4\pi a^3 \eta}{kT} \qquad (5.5)$$

(b) *Proton fluctuation* (Kirkwood and Shumaker 1952). On the surface of the protein there are acidic and basic sites; if the pH of the solution is close to the pK of any such groups then protons continually bind and disassociate, causing the dipole moment of the molecule to fluctuate with time. It may be seen in eqn (6.1) that the dielectric increment is proportional not to μ but to μ^2, and it is therefore possible to obtain a dispersion even if the average dipole moment is zero, provided there are sufficient fluctuations to provide a substantial mean square dipole moment. The origin of permittivity in the fluctuations of a system has been described in § 2.2.3 and is evident in eqn (2.18). Kirkwood and Shumaker derived the expression for the mean square dipole moment corresponding to eqn (6.23), which is discussed in § 6.3.2. The relaxation time of this process is less easy to establish and is complicated by the kinetic nature of the process.

(c) *Structured water* (Jacobson 1955). Near the surface of the protein it is certain that some restructuring of the water takes place; it is quite possible that some or all of this modified water has a permittivity higher than that of free water and that the permittivity of the whole system is therefore raised. However, most authorities would now ascribe the δ-dispersion (§ 5.4.1) rather than the β-dispersion to this effect.

(d) *Maxwell-Wagner mechanism* (Maxwell 1892; Wagner 1913, 1914). This occurs in an inhomogeneous mixture of dielectrics owing to the build up of charge on the boundaries between the different materials. One may consider the protein and the solvent to have their own conductivities and permittivities, and the β-dispersion to be caused by this mechanism. It is important with the larger biological structures and is considered in § 6.4.4.

(e) *Surface conductivity* (O'Konski 1960; Schwarz 1962). Small ions may be able to move over the protein molecule, creating additional polarization in the presence of an electric field. O'Konski adapted the Maxwell-Wagner mechanism to account for this effect and later Schwarz gave a more detailed model involving surface ion density and mobility.

(f) *Ion atmosphere* (Debye and Falkenhagen 1928; Debye and Hückel 1923). Around a protein molecule in solution there may be a cloud of small ions (rather than the very thin layer implied in (e) above) and this cloud would become polarized on the application of a field.

When investigating the β-dispersion of a protein solution, all these mechanisms should in principle be considered. However, in recent years there has been considerable evidence to show that, for reasonably small globular proteins at least, only molecular rotation and possibly proton fluctuation are of importance.

First there have been results which are in direct conflict with the predictions of the other theories. The structured water model has been shown by O'Konski (1960) to be inappropriate for Helix Pomatia hemocyanin on the grounds that it leads to impossibly high thicknesses of hydration. The linear dependence of Δ on concentration for myoglobin (Fig. 6.1) up to a fairly high concentrations is also inconsistent with this model (South and Grant 1972). The Maxwell-Wagner, surface conductivity, and ion atmosphere models all suggest that the parameters of the β-dispersion should be dependent on the conductivity of the solvent. However, there have been several studies

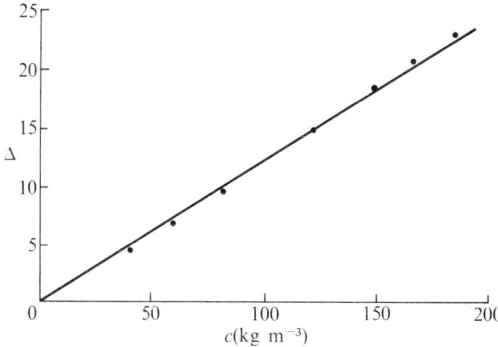

FIG. 6.1. Graph of the increment of the β-dispersion against concentration for whale myoglobin (from South and Grant 1972)

which have found no such effect; Takashima (1963) added various salts to egg albumin solutions and found no difference in the dispersion when the salts were ones which do not bind to the protein; Lumry and Yue (1965) obtained similar results for myoglobin and Moser, Squire, and O'Konski (1966), who made studies on birefringence and dielectric dispersion, were unable to find any value of surface conductivity which would explain their results on bovine serum albumin. A more detailed account of these points in the case of myoglobin is to be found in South and Grant (1972).

There are also several investigations which have verified the predictions of the molecular rotation model for the β-dispersion. An obvious test of this model is the variation of viscosity to find whether the relaxation time behaves in the way suggested by eqn (5.5). This was first done by Takashima (1962) who added glycerine to solutions of haemoglobin, myoglobin, and egg albumin and obtained linear τ against η graphs (Fig. 6.2). Similar results were obtained by Minakata (1966) on G-actin and by Hendrickx, Verbruggen, Rosseneu-Motreff, Blaton, and Peeters (1968) on bovine serum albumin. Goebel and Vogel (1964) varied the viscosity by altering the temperature and using heavy water as solvent; again the results were in agreement with eqn (5.5). South and Grant (1972) varied the

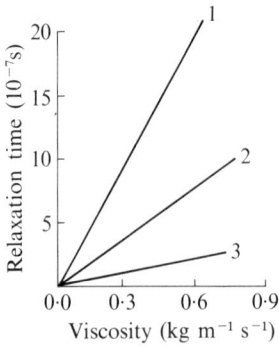

FIG. 6.2. Relaxation times of proteins in glycerin–water mixtures against viscosity of solvent 1. myoglobin, 2. egg albumin, 3. bovine haemoglobin (from Takashima 1962)

temperature of solutions of myoglobin and haemoglobin (Grant et al. 1971), and as well as confirming eqn (5.5), noticed the effect of protein–protein interactions on the relaxation time, which could be explained by the intrinsic viscosity of the protein. A similar effect was found by Schlecht, Mayer, Hettner, and Vogel (1969) for human haemoglobin.

A further test of the rotation theory is the value of molecular dipole moment to which it leads. As discussed in § 6.3.2, the experimental value for myoglobin is in good agreement with the figure computed from the molecular co-ordinates.

It appears, therefore, that deductions may be made on the basis of the rotation model, although as discussed later the presence of proton fluctuation may slightly modify these. It should also be noted that when considering structures larger than globular proteins it may well be that some of the other mechanisms listed above are of importance, and further systematic studies will be necessary to clarify this position.

6.3. The deduction of molecular parameters for proteins

6.3.1. *Size, shape, and hydration*

The physical dimensions of the protein in solution are assessed from the relaxation time (or times) of the β-dispersion, and are related to the rates of angular diffusion in the following way. If x, y, and z represent the principal orthogonal axes of the molecule with their origin at the centre of mass, then an ensemble of such molecules which have all their axes parallel at time $t = 0$ is imagined, as shown in Fig. 2.15(a). The molecules are then freed and rotate by Brownian movement, so that at a later time the orientation of the molecules is as shown in Fig. 2.15(b). This happens in such a way that if the angular density of molecules at an angle α with the original direction is n, then by the diffusion equation,

$$\frac{dn}{dt} = \theta_z \frac{d^2 n}{d\alpha^2}, \tag{6.2}$$

where θ_z is the rotary diffusion constant about the z axis.

There are two similar equations representing diffusion about the x and y axes also. This approach to rotational diffusion is the same as that employed in birefringence (Fredericq and Houssier 1973). The dielectric relaxation time of the x axis is determined by diffusion about the y and z axes, and is given by

$$\tau_x = \frac{1}{\theta_y + \theta_z} \; ; \tag{6.3}$$

there are two similar equations for the rotation of the other two axes.

Hence in general a molecule will have three relaxation times, one for each of the principal axes; but first let us consider the simplest case, that of a spherical molecule. The rotary diffusion constants will be identical ($= \theta$, say) and there will be just one relaxation time equal to $1/2\theta$, which differs from the relaxation time of $1/6\theta$ measured in birefringence studies. Values for this relaxation time will have been found by measuring the permittivity and/or the conductivity of solutions of different concentration over the β-dispersion region. The curves will have been fitted to a single relaxation time expression (§ 2.3.3) and these times extrapolated to give a value at zero concentration. There is some evidence to suggest that finite concentrations tend not only to increase the relaxation time but also to broaden the dispersion slightly, and it may therefore be found that the single relaxation time does not give a very good fit for the more concentrated solutions; however, extrapolation to infinite dilution overcomes this difficulty. This is distinguishable from the case of an asymmetrical molecule since the root mean square deviation of the single relaxation time fit should approach the experimental accuracy, which will not be so for the asymmetrical molecule.

From this relaxation time the molecular radius and volume may be calculated from eqn (5.5). As an example, Fig. 6.3 shows the variation of relaxation time of whale myoglobin with concentration at 20°C. Extrapolating to zero concentration gives a figure of 30 ns for the relaxation time. At 20°C the viscosity of water is 1.002×10^{-3} kg m^{-1} s^{-1}. Hence

$$a^3 = \frac{1.38 \times 10^{-23} \times 293 \times 30 \times 10^{-9}}{4\pi \times 1.002 \times 10^{-3}},$$

the molecular volume = $(4/3)\pi a^3 = 41 \times 10^{-27}$ m^3, and the molecular radius = a = 2.12 nm. Owing to the cube root which is taken in this calculation an error of 6 per cent in the relaxation time would only produce an error of 2 per cent in the figure for the molecular radius. More precisely, this distance is the effective radius of the rotating unit which will in general include a contribution from the thickness of any bound water on the protein surface, as was mentioned in § 5.4.3. It is assumed that viscosity of the bound water is very much greater than that of the free water and thus that the macromolecule plus water of hydration rotates as one unit at frequencies appropriate to the β-dispersion. Hence the dielectric measurements will not reveal any details about the nature of the binding but will enable the quantity of bound water present to be calculated provided that an independent assessment of the actual molecular volume is available. Macromolecules whose structure has been elucidated from the X-ray diffraction of the crystal do have known molecular volumes, and if these are not significantly different in solution then the thickness of the hydration layer may be easily found. For myoglobin the X-ray analysis reveals molecular dimensions 2.3 × 3.5 × 4 nm (Kendrew,

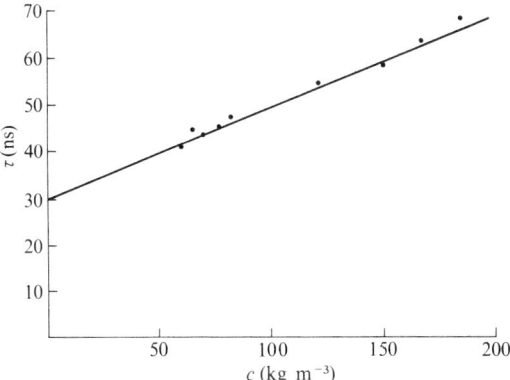

FIG. 6.3. Graph of relaxation time against concentration for a solution of whale myoglobin at 20°C (from South and Grant 1972)

Dickerson, Strandberg, Hart, and Davies 1960) which implies a volume of 17 nm^3. As was calculated above, the effective dielectric volume of the molecule is 41 nm^3 and hence a volume of 24 nm^3 of bound water is implied. If this is considered to be spread into a shell around the molecule then the width of the shell is readily found to be about 0.5 nm, i.e. about two water molecules thick.

Of course this figure is very approximate since the molecule is not perfectly spherical as has been assumed in calculating the effective radius. The effect of this molecular asymmetry may be seen from the axial ratio plot in Fig. 6.7, which indicates that if the rotating unit is assumed to be an oblate ellipsoid of axial ratio 2, as is indicated from the X-ray data of the crystal, then the hydration thickness would be 0.35 nm.

Therefore, we see that the β-dispersion results of myoglobin are consistent with the view that the molecular conformation is the same in solution as it is in the crystal provided that a hydration layer of about two molecules thickness is formed in solution. When considering the chemical and physiological behaviour of the molecule this is important since the detailed structure is found from the crystalline state in which the molecular environment is different from the natural aqueous one; it is reassuring to know that the two shapes are not very different. It is in principle possible that the molecule swells on going into solution and this could account for the relaxation time without the postulate of a hydration layer; however, this is unlikely on chemical grounds.

Let us now turn to a more general case. Many molecules are far from spherical and it is necessary to take their shape into account. As a first approximation such molecules may be considered to be ellipsoids of revolution, which includes both the prolate ellipsoid (cigar) and the oblate ellipsoid (disk). An ellipsoid with semi-axes of length a and b is imagined to be rotated about the a axis to give the three-dimensional model for the molecule; thus the prolate ellipsoid corresponds to $a > b$ and the oblate to $a < b$. The special case of $a = b$ is the sphere discussed above. Ideally two relaxation times should be found for a solution of such molecules by fitting the β-dispersion to an expression of the form

$$\varepsilon' = \frac{\Delta_a}{1 + \omega^2\tau_a^2} + \frac{\Delta_b}{1 + \omega^2\tau_b^2} + \varepsilon_\infty, \qquad (6.4)$$

where τ_a is the relaxation time of the a axis and τ_b that of the b axis. As mentioned above, it is advisable to extrapolate to zero concentration to obtain τ_a and τ_b. Owing to experimental error it is usually necessary that the two relaxation times are well separated (i.e. differ by a factor of 2 or 3 or more) to be able to obtain them with any reliability; this is due to the broad shape of the single relaxation time dispersion. The parameters Δ_a and Δ_b are proportional to the squares of the components of the dipole moment along each axis.

The relationship between the two relaxation times and the molecular size and shape has been calculated by Perrin (1934) and is given by the equations

$$\frac{\tau_a}{\tau_0} = \frac{4}{3} \frac{a^4 - b^4}{b^2(2a^2-b^2)aS - 2a^2b^2} \qquad (6.5)$$

$$\frac{\tau_b}{\tau_0} = \frac{8}{3} \frac{a^4 - b^4}{b^2(2b^2-a^2)aS + 2a^4}, \qquad (6.6)$$

where

$$S = \frac{2}{(a^2-b^2)^{1/2}} \log_e\left\{\frac{a + (a^2-b^2)^{1/2}}{b}\right\} \quad \text{if } a > b \qquad (6.7)$$

$$S = \frac{2}{(b^2-a^2)^{1/2}} \tan^{-1}\left\{\frac{(b^2-a^2)^{1/2}}{a}\right\} \quad \text{if } a < b \qquad (6.8)$$

and τ_0 is the relaxation time of a sphere of the same volume as the ellipsoid i.e. $4\pi ab^2\eta/kT$. Values of these functions are shown in Table 6.1.

If τ_a and τ_b are found, these equations enable figures to be obtained for a and b. In practice the easiest way to do this is

TABLE 6.1

Intrinsic viscosity data for ellipsoidal molecules in solution

Axial ratio	Prolate			Oblate		
	τ_a/τ_0	τ_b/τ_0	ν	τ_a/τ_0	τ_b/τ_0	ν
1	1	1	2.5	1	1	2.5
2	1.38	1.05	2.91	1.13	1.26	2.84
3	2.37	1.13	3.68	1.46	1.63	3.43
4	3.40	1.19	4.66	1.84	2.03	4.06
5	4.77	1.23	5.81	2.28	2.54	4.71
6	6.25	1.25	7.10	2.70	2.90	5.37
8	10.0	1.28	10.1	3.57	3.78	6.70
10	14.3	1.31	13.6	4.35	4.65	8.04

to calculate the ratio of the relaxation times and then to consult Fig. 6.4 to find the axial ratio, a/b. This graph is constructed on the basis of the Perrin equations and makes use of the fact that the ratio of the relaxation times depends on the axial ratio but not on the molecular volume. This graph shows that if $0.9 < \tau_b/\tau_0 < 1.0$ then a unique interpretation is not possible, since both an oblate and a prolate ellipsoid

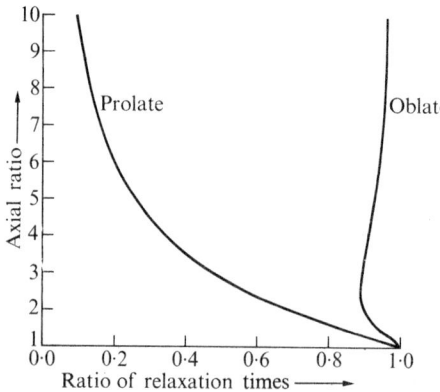

FIG. 6.4. Relationship between the ratio of relaxation times and the axial ratio for ellipsoidal molecules

can be found which are consistent with the data. Furthermore the oblate curve is extremely insensitive to the ratio of the relaxation times. However, in practice this situation is unlikely to arise since if the relaxation times are as close as this the dispersion curve will be only very slightly different from that of a single relaxation time and it is quite probable that experimental error will prevent the distinction from being made. If the molecule is behaving as a prolate ellipsoid and its axial ratio has been found in this way, the molecular volume may be calculated using Fig. 6.5, from which the ratio τ_a/τ_0 is obtained, enabling τ_0 to be found. This quantity is directly related to the molecular volume, V, by

$$V = \frac{kT\tau_0}{3\eta}. \tag{6.9}$$

Confirmation of these deductions may be made if the intrinsic viscosity $[\eta]$ of the protein is known. This is defined as

$$[\eta] = \underset{c \to 0}{\text{Limit}} \left\{ \frac{\eta_s - \eta_w}{c\eta_w} \right\}, \tag{6.10}$$

where c is the concentration of the solution, η_s is the solution viscosity, and η_w is the viscosity of water at the same temperature. This quantity is also dependent on size and shape and has been calculated by Simha (1940) to be

$$[\eta] = \frac{VL}{M} \nu, \tag{6.11}$$

where ν is a function of the axial ratio only and is shown in Table 6.1.

Scheraga (1961) and Hendrickx *et al.* (1968) have first found the rotary diffusion constants from the relaxation times using the equations

$$\theta_a = \frac{1}{\tau_b} - \frac{1}{2\tau_a} \tag{6.12}$$

$$\theta_b = \frac{1}{2\tau_a}, \tag{6.13}$$

and have then calculated the 'δ-functions' defined by

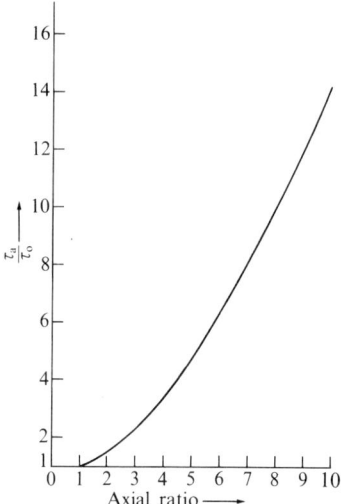

FIG. 6.5. The ratio of τ_a to τ_0 for a prolate ellipsoid

$$\delta_a \equiv \frac{6\eta_0 \theta_a [\eta] M}{LkT} \qquad (6.14)$$

$$\delta_b \equiv \frac{6\eta_0 \theta_b [\eta] M}{LkT} \,. \qquad (6.15)$$

These are dependent on the axial ratio only and may be applied to Fig. 6.6 to find the most appropriate axial ratio; this should agree with the figure found by the first method described above.

As an example of these calculations we may take the figures of Moser et al. (1966) for bovine serum albumin. By use of birefringence and dielectric measurements, they found the two relaxation times to be τ_a = 0.23 μs and τ_b = 0.11 μs at 25°C. Thus the ratio of the relaxation times is 0.48, and from Fig. 6.4 it may be seen that this corresponds to a prolate ellipsoid with an axial ratio of about 2.9. This figure may be checked by calculating the δ-functions; from the relaxation times and eqns (6.12) and (6.13) it is found that θ_a = 7.8×10^6 s^{-1} and θ_b = 2.18×10^6 s^{-1}. Since the intrinsic viscosity of BSA is 4.3×10^{-5} m^3 kg^{-1}, eqns (6.14) and (6.15)

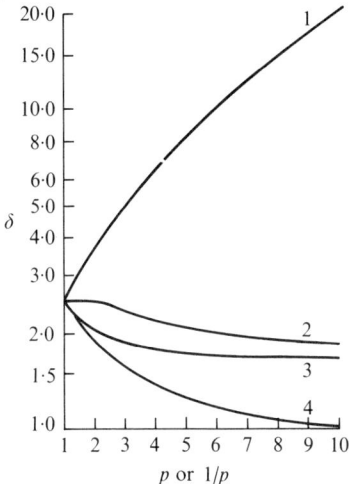

FIG. 6.6. Dependence of δ_a and δ_b on the axial ratio p for prolate and oblate ellipsoids: curve 1, δ_a, prolate; curve 2, δ_b, oblate; curve 3, δ_a, oblate; curve 4, δ_b, prolate (from Hendrickx *et al.* 1968)

give $\delta_a = 5.0$ and $\delta_b = 1.6$. From Fig. 6.6 it is seen that these also imply an axial ratio of about 3.

The molecular volume and dimensions may now be found, since from Fig. 6.5 it is seen that for an axial ratio of 3, $\tau_a/\tau_0 = 2.4$. Hence $\tau_0 = 0.23/2.4$ µs = 0.097 µs, and from eqn (6.9)

$$V = \frac{1.38 \times 10^{-23} \times 298 \times 0.097 \times 10^{-9}}{3 \times 0.92 \times 10^{-3}}$$

$$= 1.5 \times 10^{-25} \text{ m}^3.$$

a and b may now be found since $V = (4/3)\pi ab^2 = (4/3)\pi 3b^3$. Therefore

$$b^3 = V/4\pi$$

$$= 11.9 \times 10^{-27} \text{ m}^3$$

and hence

$$b = 2.3 \times 10^{-9} \text{ m},$$

from which it follows that a (= $3b$) is 6.9×10^{-9} m.

It sometimes happens in practice that because the two relaxation times are very close, as for the oblate or nearly spherical molecule, or if the dipole moment lies predominantly along one axis, then only one relaxation time may be discerned from the dispersion. In this case no unique interpretation is possible unless some additional information is available. For instance, the effective molecular radius may be calculated from eqn (6.2) and it may be that the molecule really is spherical, but on the other hand it may be that the molecule is really ellipsoidal and the observed relaxation time is that about just one of the axes; there are then many different shapes and sizes which would all be in accord with the observations. The possibilities may be displayed graphically, as is done for myoglobin in Fig. 6.7. This plot may be constructed from the Perrin functions in the following way. The molecular volume corresponding to an axial ratio of one (i.e. the sphere) is

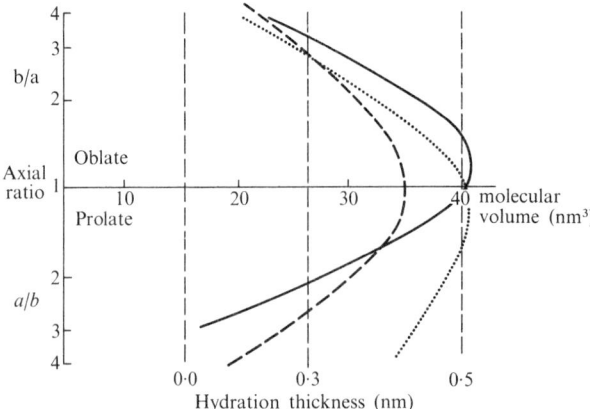

FIG. 6.7. Plot of axial ratio against molecular volume for myoglobin. Solid curve, dielectric relaxation, a-axis; dotted curve, dielectric relaxation, b-axis; dashed curve, viscosity measurement (from South and Grant 1972)

calculated from eqn (6.9) where the observed relaxation time is put in the place of τ_0. If this volume is V_0 and the observed relaxation time of the molecule is thought to be that of the a axis then

$$V = \frac{kT\tau_0}{3\eta} \qquad (6.16)$$

and

$$V_0 = \frac{kT\tau_a}{3\eta}, \qquad (6.17)$$

which taken together give

$$\frac{V}{V_0} = \frac{\tau_0}{\tau_a} = 1 \bigg/ \frac{\tau_a}{\tau_0}. \qquad (6.18)$$

τ_a/τ_0 is given as a function of the axial ratio in Fig. 6.5, and hence this equation is that of the a-axis line in Fig. 6.7. A similar expression concerning the b-axis gives the equation of the other line. Hence in principle the molecule may have a configuration represented by a point on either of these lines, or if the measured relaxation time is thought to be some average of the two real relaxation times then the molecules' size and shape will be represented by a point somewhere between the two lines, the exact position depending on the weighting of the average. If either the axial ratio or the molecular volume is known independently then a more exact conclusion may be made; an example of this is the use of the axial ratio given by X-ray diffraction of the crystal which is discussed in the next section.

An alternative means of supplying additional information is the use of the intrinsic viscosity mentioned above. Eqn (6.11) is also the equation of a line if axial ratio is plotted against molecular volume, as in Fig. 6.7, since ν is a function of axial ratio only. This line is shown in the graph and has been calculated using a value of 3.1×10^{-3} m^3 kg^{-1} (Wyman and Ingals 1943). In principle the correct configuration is given by the point where this line crosses the appropriate relaxation time line, so that in this case a possible, but not unique, configuration is that of a prolate molecule with axial ratio 1.5 and volume 34×10^{-27} m^3. The errors in these deductions may be assessed by considering how much the lines would move if different figures were used for the relaxation times and the intrinsic viscosity.

It would be a natural progression to consider now molecules as ellipsoids with three different diameters, but as yet no measurements have been made with sufficient accuracy to enable such an analysis to be performed and there is therefore no purpose in discussing this case here. However, it is very much to be hoped that in due course apparatus will be refined to such an extent as to make this possible, for it would greatly increase the power of dielectric studies in the discovery of macromolecular shape.

Before leaving this topic of the β-dispersion relaxation time, it is worth noting that useful deductions can sometimes be made from such quantities without the detailed analysis described above. Such is the case when merely comparing the configuration of two molecules as has been done for various forms of haemoglobin. This was first done by Takashima and Lumry (1958*a*, *b*) who varied the pressure of oxygen and carbon monoxide above haemoglobin and detected distinct maxima and minima in the relaxation times as the pressure was changed, which at the time was felt to be due to changes in the conformation of the protein in the changing environment. However, recent studies (Hanss and Bannerjee 1967; Schlecht, Vogel, and Mayer 1968; Grant, South, Takashima, and Ichimura 1971) have failed to confirm this effect; it now seems that the original work was in error and that the conformation of haemoglobin does not change significantly when the bound ligand is changed.

Another study using dielectric measurements to investigate protein conformation is that of Rosseneu-Motreff, Soetewey, Lamote, and Peeters (1973). The denaturation of delipidated and relipidated BSA by urea has been monitored by analysis of the relaxation time and dielectric increment; they find that the denaturation causes both an elongation and a swelling of the molecule and also that BSA is less susceptible to denaturation when lipidated rather than delipidated.

Dielectric measurements are also of potential use in studying chemical rate processes when the reaction changes the dipole moment of the system, e.g.

$$A_1 \underset{k_{21}}{\overset{k_{12}}{\rightleftharpoons}} A_2 .$$

If A_1 has no dipole moment but A_2 does, then the application of a field will favour the production of A_2 oriented in the direction of the field. We will not discuss this further here, but more detailed considerations are given by Schwarz (1972).

6.3.2. *Dipole moments*

As has been discussed in § 6.2, the β-dispersion of globular proteins is created by the rotation of the molecules which possess a dipole moment. It has also been mentioned that the dielectric increment, Δ, of the β-dispersion is generally proportional to the concentration of the solution, c, and hence the specific increment, $\delta (= \Delta/c)$, is a well-defined quality and it is this parameter that is related to the molecular dipole moment. In order to establish the precise nature of this relationship some model must be used to represent the solution and it has been usual to assume that both the water and macromolecules behave as point dipoles and therefore that the theoretical approach of § 2.3.4 may be applied; in particular eqns (2.66) and (2.67) describe the dispersion. Eqn (2.67) may be modified to eqn (6.1) in order to express $g_A \mu_A$ in terms of directly measurable quantities. The value of g_A is generally not known and it has been common to calculate a molecular dipole, μ_{exp}, using the formula

$$\mu_{exp} = \left(\frac{2\varepsilon_0 kTM\delta}{L} \right)^{\frac{1}{2}} , \qquad (6.19)$$

i.e. it is assumed that g_A is unity. This is probably a reasonable assumption since the macromolecule has a large surface area and any local regions around the molecule where short-range forces create dipole moments are likely to cancel each other when averaged over the entire molecule; in an extreme case where short-range forces are considered to be spherically symmetrical, g_A would be exactly unity. It is therefore usually values of μ_{exp} which are quoted in the literature but it should be recalled that, following the approach used in relation to small biological molecules (§ 5.5.2),

$$\mu_{exp} = g^{\frac{1}{2}}\mu_A. \qquad (6.20)$$

Dipole moments calculated in this way do not refer to an isolated molecule (i.e. in an imaginary gas phase), since the dipole of the latter includes a polarization induced by the cavity field within the molecule; μ_A refers to the dipole arising from the distribution of permanent charge throughout the molecule. The relationship between these two dipoles, in the case of a spherical molecule, has been given by eqn (2.23). It should also be borne in mind that all these equations have been based on a model of point dipoles which is only an approximation to the real situation in which the molecules have finite size and have their dipole moments distributed over their volume. An indication that eqns (2.67) and (6.1) are sensitive to the model has been given by considering the water as a continuum with its known macroscopic parameter, rather than as a set of discrete dipoles (South and Grant 1974). In this approximation the factor of ½ in these equations becomes ¾ and hence leads to a value of μ_{exp} smaller by a factor of about 0.82 than given by eqn (6.19). All these models have their limitation and it is to be hoped that eventually a more exact theory will be developed, possibly from the method of computer simulation described earlier for the case of pure water. Meanwhile values of experimental dipole moment quoted in this chapter are calculated from eqn (6.19). For small proteins typical values of dipole moment are several hundred Debye units, which corresponds to one or two electrons displaced by the molecular diameter. This is the dipole which would be obtained by removing one electron from a homogeneous molecule, with zero dipole moment, and replacing it on the other side of the molecule. Thus the value of the observed dipole moment reflects the high degree of electrical symmetry in protein molecules.

In practice, of course, the distribution of charge is very

much more complicated than this, but if the distribution were known then the expected dipole could be calculated and its value compared with the experimental figure. The purposes of such a comparison would be (a) to test the dipole rotation theory, (b) to investigate whether the dipole moment was the same in solution as it is in the crystal, and (c) to confirm the value of the dipole moment which may be of significance in the physiological behaviour of the molecule.

The success of X-ray diffraction in establishing the structure of some proteins has now made such an investigation possible and this line of attack has been followed in the case of myoglobin by Schlecht (1969), by Grant, South, and Walker (1971), and by South and Grant (1972).

The calculation of the exact dipole moment from the atomic coordinates would be an immense task since the contribution from every bond would have to be included. However, an approximate figure can be calculated if only the dominant causes of the dipole moment are considered; of these the most important appears to be the effect of the charged groups on the surface of the molecule, i.e. those sites which are able to lose or gain a proton and become negatively or positively charged. Thus for the molecule shown in Fig. 6.8 the dipole moment from this effect would be

$$\mu = e\mathbf{r}_1 + e\mathbf{r}_2 - e\mathbf{r}_3 - e\mathbf{r}_4 , \qquad (6.21)$$

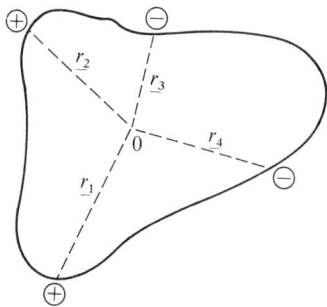

FIG. 6.8. Simplified model of a protein. The dotted lines represent vectors and the point 0 is the centre of mass

where r_1, r_2, etc. represent the position vectors of the charges In this particular case the origin of the position vectors is irrelevant since the molecule has no net charge; however, if this is not the case and the molecule does carry a net charge then the choice of origin would effect the calculated view of the dipole. The centre of mass of the molecule is the correct choice of origin, since it is about this point that dielectric rotation takes place.

Whether or not an ionizable site is charged depends on the pH of the solution. An acid site is negatively charged at values of pH well above the pK value and a basic site is positively charged at pHs well below its pK value. At pHs around the pK value the site may or may not be charged, and from the law of mass action the probability of a site carrying a proton is given by

$$f_i(\text{pH}) = \frac{1}{1 + 10^{\text{pH} - \text{p}K_i}}, \tag{6.22}$$

the subscript referring to the ith site. Thus the time or ensemble average charge on a site is given by

$$e\{f_i(\text{pH}) - \delta_i\},$$

where δ_i is 0 when site i is basic and 1 when it is acidic. The contribution of the surface charges to the dipole moment may therefore be written

$$e \sum_{i=1}^{n} \{f_i(\text{pH}) - \delta_i\} r_i,$$

where the sum is taken over all the ionizable sites on the molecule, and r_i is the position vector of the ith site. This calculation is performed by computer and the results for myoglobin are shown in Fig. 6.9, where it may be seen immediately that the values obtained are of the same order of magnitude as the experimental figures, and that the dipole moment of horse myoglobin is 20–30 D greater than that of whale myoglobin. The core of the protein will also contribute to the dipole moment and the magnitude of this effect may be assessed by using the estimate of Wada (1959, 1960) that the dipole moment per amino acid along an α-helix is 3.5 D. This contribution has also been included in Fig. 6.9. This is only an approximation to the contribution from the core of the protein and further refinements would include the effects from the bonds joining the amino acid to the peptide chain and also the bending and consequent deformation of the chain.

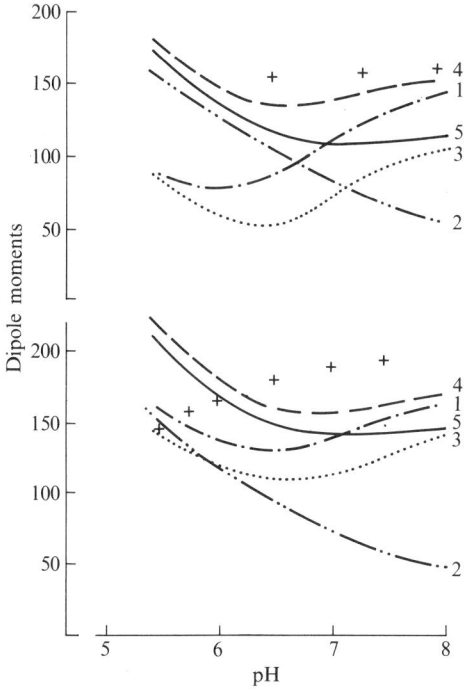

FIG. 6.9. Calculated and experimental dipole moments for whale and horse myoglobin as a function of pH. Crosses are experimental values. 1, surface charges only; 2, fluctuation dipole only; 3, surface charges + core dipole; 4, surface charges + fluctuation dipole; 5, surface charges + fluctuation dipole + core dipole (from South and Grant 1972)

Another contribution to the molecular dipole moment may arise from the fluctuation effect discussed by Kirkwood and Shumaker (1952). These authors were able to make only a rough estimate of the magnitude of the mean square fluctuation moment, but now that molecular coordinates are known a more accurate figure may be computed from the expression

$$\overline{\Delta\mu^2} = e^2 \sum_{i=1}^{n} (f_i(1-f_i))r_i^2 , \qquad (6.23)$$

which may be proved by averaging the square of the dipole moment (South 1970). $\overline{\Delta\mu^2}$ is the mean square fluctuation moment and is related to the mean square moment, $\overline{\mu^2}$ and the mean moment, $\overline{\mu}$ by

$$\overline{\mu^2} = \overline{\mu}^2 + \overline{\Delta\mu^2} . \qquad (6.24)$$

This contribution is also shown in Fig. 6.9 where it is seen that the various computed dipole moments are all fairly close to the experimental figures and it is not possible to say at the moment which curve is the most realistic. However, some light has been thrown on the question of the proton fluctuation contribution (Scheider 1965; South and Grant 1973). In general a protein with the structure outlined above will show dispersion caused by rotation and also by proton fluctuation. The relaxation time of the latter, τ_p, is given by

$$\frac{1}{\tau_p} = \frac{1}{\tau} + \frac{1}{\tau'} , \qquad (6.25)$$

where τ and τ' are the rotational and fluctuational correlation times respectively. This equation is a consequence of the fact that the dipole created by the proton positions can decay by either rotation or proton jumping. It follows that if $\tau' \ll \tau$, then $\tau_p \ll \tau$ and the two dispersions will be quite distinct; on the other hand if $\tau' \gg \tau$, then $\tau_p \approx \tau$ and the dispersions will overlap and probably be indistinguishable, giving rise to one dispersion created by an effective dipole moment of $(\overline{\mu}^2+\overline{\Delta\mu^2})^{\frac{1}{2}}$.

Two sources of error are immediately apparent in these calculations; first, particular values have been assumed for the pK of each group and it is possible that these are not quite correct, indeed it may be that two amino acids of the same type have slightly different pKs depending on the exact location of the group in the molecule. Secondly, the coordinates are not known precisely for some of the important atoms since they are only loosely bound to the main chain and are partially mobile in solution; the relatively good agreement between the experimental and computed dipoles would seem to indicate that these sources of error are not of great importance.

A slightly different approach has been taken by Schlecht (1969) who added carboxymethyl groups to the myoglobin surface, thus changing the isoelectric point of the molecule, and measured the dipole moment at various stages of binding. He then calculated the permanent dipole moment of the modified molecules in a similar fashion to the calculations above, thus avoiding the use of a particular origin since the molecule is always uncharged. He did not take into account proton fluctuations or the contribution from the α-helix but nevertheless obtained excellent agreement between calculated and experimental values. The good agreement between experimentally and theoretically calculated dipole moments found in independent studies serves to justify the various assumptions which have had to be made, and also indicate that other factors which may be thought to be present (for example, higher order electrical moments, surface saturation effects) are not of major importance. The use of dipole moments therefore seems to offer important information in macromolecular structure and is also of potential value in the investigation of the interaction and binding of two biological molecules.

6.4. Larger structures
6.4.1. *DNA*

We turn now from molecules of molecular weight of tens of thousands to DNA, whose molecular weight may be several million and whose structure is that of the rod-like double helix, in contrast with the globular proteins discussed above. These considerable differences in structure mean that the mechanisms discussed, and largely rejected, in connection with proteins (§ 6.2) may now be more relevant and that effects other than rotation of dipoles may be present.

There are at least two dispersion regions in aqueous DNA solutions, one having a relaxation time of about 1 ms and the other of about 100 ns. The lower frequency dispersion has been studied by Hanss (1966) and by Takashima (1966, 1967a, b) who found very high dielectric increments (e.g. 1800 for a solution of concentration of 0.1 kg m^{-3}) which appear to be proportional to the square of the length of the molecule; the relaxation time also has this same dependence on DNA length, but its value is not in good agreement with the rotary diffusion constant as found by birefringence studies. Takashima believed this dispersion to be due to counterions in the system (a mechanism of the type envisaged by Schwarz (1962)) and he added various small ions to the solution. The addition of sodium chloride and magnesium chloride had the effect of greatly reducing the increment and the relaxation time, although the dimensions of the molecule changed only slightly. This may be explained qualitatively by a model in which binding sites, separated from each other by potential barriers of constant height, are placed along the length of the rigid molecule. The dipole moment is then dependent on the number of counterions which are bound to the molecule, and the relaxation time is dependent on the mobility of the counterions. A similar but more detailed theory has been advanced by Minakata, Imai, and Oosawa (1972) to explain this dispersion in terms of the counterion fluctuation along the length of the molecule. Takashima (1973) has shown experimentally that the dipole moment is directed predominantly along the long axis of the molecule and also that it is an induced rather than a permanent moment. This was done by using a dielectric cell with two concentric cylinder electrodes, one of which was able to rotate and thus align the molecules at right angles to the applied field; the permittivity fell with increased alignment, which is consistent with the conclusions just mentioned.

All this evidence suggests that this low-frequency dispersion is caused by the fluctuation of counterions along the length of the molecule. Other approaches to the theory of this effect have been given by Mandel (1961, 1977), McTague and Gibbs (1966), and by Oosawa (1970). In another study Hanss and Bernengo (1973) measured the conductivity dispersion of samples of calf thymus DNA and found a low-frequency dispersion which was analysed into two components with relaxation times of typically 2 ms and 20 ms. These values agreed with those found from measurements of reversing pulse birefringence and were thought to be explicable only in terms of an orientation mechanism, a dipole moment possibly being created by the difference in ionic character of the two molecular ends.

Takashima (1965) has used this dispersion to study the helix-coil transition in DNA; he finds that the dipole moment and the relaxation time are greatly reduced in the coil form, which is consistent with the counterion theory mentioned above. There also seems to be some evidence in the dispersion curves for an intermediate form of DNA.

Another application of this dispersion is to the study of DNA-proflavine complexes (Goswami, Das, and Das Gupta 1973). It was found that with increased binding of proflavine the permittivity decreases and the relaxation time increases. This is interpreted in terms of an increase in both the length of the molecule and the neutralization of its surface charges by the proflavine.

Neale and Weyl (1966) have made measurements largely of the higher frequency dispersion for which they confirmed the findings of Junger, Junger, and Allgen (1949) that DNA solutions possess a dispersion region around a few MHz whose relaxation time decreases as the concentration of the solution is increased. Also Allgen (1954) found that there was little variation in the dispersion when deaminated DNA was used. These facts are certainly at variance with the model of molecular rotation and it seems possible that the dispersion is caused by the O'Konski (1960) mechanism of ionic movement on the molecular surface. The experimental data are consistent with a figure of around $\Omega^{-1}\,m^{-1}$ for the surface conductivity in the direction of the long axis of the molecule (Neale and Weyl 1966).

6.4.2. *Lipoproteins*

The permittivity of aqueous solutions of human and bovine low density lipoproteins (LDL) has been measured by Essex *et al.* (1977*b*) between 0.1 and 1000 MHz and at 800 MHz only by Grant *et al.* (1972). One purpose of these investigations (briefly referred to in § 5.4.3) was to compare the values of

δ_{800} between normal samples and those arising from various pathological states. In these investigations the permittivity of the solutions of LDL taken from patients with familial hyperbetalipoproteinaemia (type II) was found to exhibit a slightly lower value than control samples of LDL. This effect was more marked for homozygote patients than for heterozygotes and may be interpreted in terms of a greater quantity of bound water in the LDL of the patients since such water will exhibit a fairly low permittivity at 800 MHz (§ 5.4.2). This interpretation is also consistent with the results of ultracentrifuge measurements, but further work will be necessary before any firm and final conclusions are drawn.

The lower frequency measurements of Essex *et al.* were interpreted in terms of an α-dispersion with a relaxation frequency around 0.5 MHz and a β-dispersion having a relaxation frequency near 5 MHz. The α-dispersion was attributed to the relaxation of counterions and the β-dispersion to a Maxwell-Wagner type mechanism as described by Pauly and Schwan (1966). The measurements and their interpretation favour the lipid bilayer model, as against the lipid core model, and the non-linearity of the increment due to the α-dispersion strongly suggested the formation of micelles or aggregates. The concentration threshold for this non-linearity was very low; 20 mg/ml for bovine LDL and 30 mg/ml for human LDL.

6.4.3. *Viruses*

Measurements on alfalfa mosaic virus have been made (Van der Touw, Briedé, and Mandel 1973) at frequencies around 1 MHz and a well-defined dispersion discovered with a relaxation time around 600 ns and a specific increment of 2 ml/mg which was also consistent with a Cole—Cole distribution parameter of 0.6. The authors show that these data are inconsistent with the predictions of the rotational or O'Konski theories but are well explained by the counterion theory of Mandel (1961) and Oosawa (1970). Further measurements on structures such as these should yield interesting information, particularly on their surface properties.

6.4.4. *Cells and membranes*

The application of electrical measurements and concepts to cell structure has a long and very interesting history which cannot be adequately described here; the interested reader may consult K.S. Cole (1965). We give here only a brief review of some of the dielectric measurements made on a variety of cell-like structures.

The basic model behind such measurements is that shown in Fig. 6.10. The suspending medium, the cell wall, and the cell interior each have their own characteristic permittivity, $\hat{\varepsilon}_1$, $\hat{\varepsilon}_2$, $\hat{\varepsilon}_3$ which are complex in order to include the conductivities of each component, i.e. $\hat{\varepsilon}_j = \varepsilon'_j - i S_j/\omega$ where $j = 1, 2, 3$, ε'_j is the real part of the permittivity, and S_j is the conductivity (divided by ε_0). The equations which relate these complex permittivities to that of the whole suspension of cells have been developed by Fricke (1925, 1955) and may be written

$$\frac{\hat{\varepsilon} - \hat{\varepsilon}_1}{\hat{\varepsilon} + x\hat{\varepsilon}_1} = p \frac{\hat{\varepsilon}_e - \hat{\varepsilon}_1}{\hat{\varepsilon}_e + x\hat{\varepsilon}_1}, \tag{6.26}$$

where p is the volume fraction of cells and

$$\frac{\hat{\varepsilon}_e - \hat{\varepsilon}_2}{\hat{\varepsilon}_e + x\hat{\varepsilon}_2} = \left(1 + \frac{d}{R}\right)^3 \frac{\hat{\varepsilon}_3 - \hat{\varepsilon}_2}{\hat{\varepsilon}_3 + x\hat{\varepsilon}_2}. \tag{6.27}$$

In these equations $\hat{\varepsilon}_e$ is the effective permittivity of the cell if it were homogeneous and the factor x depends on the shape of the cells and is 2 in the often-assumed case of a sphere. In any situation where a mixture of two dielectric materials occurs the overall complex permittivity is frequency dependent in a manner which depends on the permittivity and conductivity of each component. This is an example of the Maxwell-Wagner effect.

The manipulation of these equations can become rather involved, although it is simple enough by computer. If computing is not possible some simplification is often helpful, a particularly useful approximation being given by Pauly, Packer, and Schwan (1960) for the case of $d \ll R$ and $S_2 \ll S_1, S_3$. A dispersion with a single relaxation time is then predicted if the parameters are frequency independent; the relaxation time, τ, and the low-frequency permittivity, ε_s, are then given by

$$\tau \approx RC \frac{S_3 + 2S_1}{2S_3 S_1} \qquad (6.28)$$

$$\varepsilon_s \approx \varepsilon_1 + \frac{9}{4\varepsilon_0} RCp , \qquad (6.29)$$

where C is the capacitance per unit area of the wall, i.e. using eqn (2.7),

$$C = \frac{\varepsilon_2 \varepsilon_0}{d} . \qquad (6.30)$$

The value of ε_2 is likely to be 10–20 (Schwan and Cole 1960) and hence measurement of C provides an estimate of the wall thickness, d. Measurement of C is best made from eqn (6.29) since this requires no prior knowledge of the conductivities. Over a wide range of membranes C is found to lie in the range 0.5–2 μf/cm^2 which (eqn (6.30)) implies $d \approx 10$ nm, a figure which is consistent with the lipid bilayer model of the cell wall.

If the relaxation time and S_1 are measured, then eqn (6.28) may be used to estimate S_3, the conductivity of the cell's interior; this is of interest since S_3 is governed by the number of ions within the cell and their mobility.

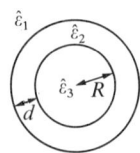

FIG. 6.10. Simple representation of a cell

One of the simplest applications of this approach is to phospholipid vesicles (Schwan, Takashima, Miyamoto, and Stoeckenius 1970). Two dispersions are found; the lower frequency one (1 kHz–1 MHz) is caused by counterion movement and is therefore not predicted directly by the model above. The other (1 MHz–100 MHz) may be described by the method above; it yields $C \approx 2$ μf/cm^2 and indicates that the internal material has very similar electrical properties to the suspending liquid. Similar studies on mitochondria (Pauly, Packer and Schwan 1960; Pauly and Packer 1960) and on yeast cells (Asami, Hanai, and Koizumi 1976) give $C \approx 1$ μf/cm^2.

Measurements on erythrocyte suspensions and whole blood have been performed for many years on a variety of animal as well as human samples (Höber 1913; Fricke and Morse 1926; Rajewsky and Schwan 1948; Fricke and Curtis, 1934). Investigations at microwave frequencies only were carried out by Cook (1952) and by Grant and Sheppard (1974b). The general picture arising from these measurements is that the permittivity of whole blood falls from a value of a few thousand at low radio frequencies to around 60 at several hundred MHz, from which it then falls to a value between 4 and 5 as the frequency increases through the microwave region to several hundred GHz. Pauly and Schwan (1966) show that a value of $\varepsilon_e \approx 60$ near 250 MHz is what would be expected for an aqueous haemoglobin solution of the equivalent concentration. They also found that the value of S_3 is only about a half of the value to be expected from the known salt content of the cytoplasm. A similar situation has been discovered in nerve fibre (Hodgkin and Keynes 1953), in lens material (Pauly and Schwan 1964) and in liver cells (Schwan and Li 1953). Thus it appears that the ionic mobility is reduced by the electrostatic and hydrodynamic interactions within the cell.

There has also been a series of measurements on bacteria, particularly $E.$ $coli$ and micrococcus (Fricke, Schwan, Li, and Bryson 1956; Carstensen, Cox, Mercer and Natale 1965; Einolf and Carstensen 1969, 1973; Carstensen and Marquis 1968; Carstensen 1967). The model of Fig. 6.10 was again used but the presence of a poorly conducting cytoplasmic membrane was represented by making $S_3 = 0$. The low-frequency electrical properties were found to reflect the presence of the cell wall which appeared to be highly conducting as a result of a large fixed charge density in this region.

BIBLIOGRAPHY

AARON, M.W. (1966). Ph.D. Thesis, University of London.
AARON, M.W. and GRANT, E.H. (1963). *Trans. Faraday Soc.* **59**, 85.
AARON, M.W. and GRANT, E.H. (1967a). *Trans. Faraday Soc.* **9**, 2177.
AARON, M.W. and GRANT, E.H. (1967b). *Br. J. appl. Phys.* **18**, 957.
ADEY, W.R. and BAWIN, S.M. (1977). Conference Brain interactions with weak fields, published as *Neurosciences Res. Prog. Bull.* **15**.
ALLGEN, L.G. (1954). *Biochim. biophys. Acta* **13**, 448.
ALLISON, J., BENSON, F.A., and SEAMAN, M.S. (1957). *Proc. Instn elect. Engrs* **104-(B)**, 599.
ASAMI, K., HANAI, T. and KOIZUMI, N. (1976). *J. Membrane Biol.* **28**, 169.
ASFAR, M.N. and HASTED, J.B. (1977). *J. Opt. Soc. Am.* **67**, 902.
BADEN FULLER, A.J. (1969). *Microwaves*, Pergamon, Oxford.
BARBENZA, G.H. (1969). *J. Phys. E.* **2**, 871.
BASSETT, H.L. (1971). *Rev. scient. Instrum.* **42**, 200.
BATEMAN, J.B. and POTAPENKO, G. (1940). *Phys. Rev.* **57**, 1185.
BATHER, R., WEBB, S.J., and CUNNINGHAM, T.A. (1965). *Nature, Lond.* **207**, 30.
BENSON, F.A. (1953). *Proc. Instn elect. Engrs* **100(111)**, 85.
BERBERIAN, J.G. and COLE, R.H. (1969). *Rev. scient. Instrum.* **40**, 811.
BERENDSEN, H.J.C. and MIGCHELSEN, C. (1965). *Ann. N.Y. Acad. Sci.* **125**, 365.
BERGMANN, K., ROBERTI, D.M., and SMYTH, C.P. (1960). *J. phys. Chem., Ithaca* **64**, 665.
BERNAL, J.D. (1964). *Proc. R. Soc.* A **280**, 299.
BERNAL, J.D. and FOWLER, R.H. (1933). *J. chem. Phys.* **1**, 515.
BJERRUM, N. (1951). *K. danske Vidensk. Selsk. Skv.* **27**, 1.
BLEANEY, B.I. and BLEANEY, B. (1976). *Electricity and magnetism* (2nd ed.), Clarendon Press, Oxford.
BOTTREAU, A.-M., DELBOS, G., MARZAT, C., LACROIX, Y., SALEFRAN, J.-L., and DUTUIT, Y. (1973). *C.r. hebd. Séanc. Acad. Sci., Paris* **276**, 373.
BOTTREAU, A.-M., DELBOS, G., MARZAT, C., and SALEFRAN, J.-L. (1973). *C.r. hebd. Séanc. Acad. Sci., Paris* **277**, 639.
BOTTREAU, A.-M., DELBOS, G., MARZAT, C., SALEFRAN, J.-L., and MOREAU, J.-M. (1974). *C.r. hebd. Séanc. Acad. Sci., Paris* **278**, 767.
BREHM, G.A. and STOCKMAYER, W.H. (1973). *J. phys. Chem., Ithaca* **77**, 1348.
BREY, W.S., HEEB, M.A., and WARD, T.M. (1969). *J. Colloid Interface Sci.* **30**, 13.
BROADHURST, M.G. and BUR, A.J. (1965). *J. Res. natn Bur. Stand.* **690**, 165.
BROPHY, J.J. and WEBB, S.L. (1962). *Phys. Rev.* **128**, 584.
BROWNLEE, K.A. (1965). *Statistical theory and methodology in science and engineering*, Wiley, New York.
BUCCI, O.M., CORTUCCI, G., FRANCESCHETTI, G., SAVARESE, C., and TIBERIE, R. (1972). *IEEE. Trans.* **IM-21**, 237.
BUCHANAN, T.J. (1952). *Proc. Inst. Radio Engrs* **99**, III, 61.
BUCHANAN, T.J. and GRANT, E.H. (1955). *Br. J. appl. Phys.* **6**, 64.
BUCHANAN, T.J., HAGGIS, G.H., HASTED, J.B., and ROBINSON, B.G. (1952). *Proc. R. Soc.* A**213**, 379.
BUCKINGHAM, A.D. (1953). *Aust. J. Chem.* **6**, 323.
BURGER, M.M. and NOONAN, K.D. (1970). *Nature, Lond.* **228**, 512.
CALVERT, R. (1948). *Wayne Kerr Monograph No. 1*. Wayne Kerr Company, Bognor Regis, U.K.
CARSON, J.W. (1970). *IEEE. Trans.* **MTT-18**, No. 1, 57.
CARSTENSEN, E.L. (1967). *Biophys. J.* **7**, 493.
CARSTENSEN, E.L., COX, M.A., MERCER, W.B., and NATALE, L.A. (1965). *Bio-*

phys. J. **5**, 289.
CARSTENSEN, E.L. and MARQUIS, R.E. (1968). *Biophys. J.* **8**, 536.
CATCHPOOL, J.F. (1965). *Ann. N.Y. Acad. Sci.* **125**, 595.
CHAMBERLAIN, J., HAIGH, J., and HINE, M.J. (1971). *Infrared phys.* **11**, 75.
CHISHOLM, J.S.R. and MORRIS, R.M. (1966). *Mathematical methods in physics*, p. 546. North-Holland Publishing Company, Amsterdam.
CLARK, A.H., QUICKENDEN, P.A., and SUGGETT, A. (1974). *J. chem. Soc. Farad. Trans. II* **70**, 1847.
CLARKSON, T.S., GLASSER, L., TUXWORTH, R.W., and WILLIAMS, G. (1977). *Adv. Mol. Relax. Proc.* **10**, 173.
CLEGG, J.S. (1978). *Proceeding of 1st International Conference on Cell Biology* (ed. W. Drost-Hansen). Academic Press, London.
COHN, E.J. and EDSALL, J.T. (1943). *Proteins, amino-acids and peptides.* Reinhold, New York.
COLE, K.S. (1965). *Theoretical and mathematical biology* (ed. T.H. Wateman and H.J. Morowitz). Blaisdell Publ. Comp., Waltham, Mass.
COLE, K.S. and COLE, R.H. (1941). *J. chem. Phys.* **9**, 341.
COLE, R.H. (1965). *J. chem. Phys.* **42**, 637.
COLE, R.H. (1974). *J. phys. Chem., Ithaca* **78**, 1440.
COLE, R.H. (1975a). *J. phys. Chem., Ithaca* **79**, 93.
COLE, R.H. (1975b). *J. phys. Chem., Ithaca* **79**, 1459.
COLE, R.H. (1975c). *J. phys. Chem., Ithaca* **79**, 1469.
COLE, R.H. and GROSS, P.M. Jr (1949). *Rev. scient. Instrum.* **20**, 252.
COLLIE, C.H., HASTED, J.B., and RITSON, D.M. (1948). *Proc. phys. Soc.* **60**, 71.
CONNER, W.P., CLARKE, R.P., and SMYTH, C.P. (1942). *J. Am. chem. Soc.* **64**, 1379.
CONNER, W.P. and SMYTH, C.P. (1942). *J. Am. chem. Soc.* **64**, 1870.
COOK, H.F. (1951). *Br. J. appl. Phys.* **2**, 295.
COOK, H.F. (1952). *Br. J. appl. Phys.* **3**, 249.
COOLEY, J.W. and TUKEY, J.W. (1965). *Maths Comput.* **19**, 297.
COULSON, C.A. and EISENBERG, D. (1966). *Proc. R. Soc.* **A291**, 445.
CROOM, E.J. (1973). Private communication.
CROOM, E.J., SHACK, R., SHEPHERD, J.C.W., and SHEPPARD, R.J. (1978). *J. Microwave Power* **12**, 111.
CULLEN, A.L., NAGENTHIRAM, P., and WILLIAMS, A.D. (1972). *Proc. R. Soc.* **A329**, 153.
CULLEN, A.L. and YU, P.K. (1971). *Proc. R. Soc.* **A325**, 493.
CUTNELL, J.D., KRANBUEHL, D.E., TURNER, E.M., and VAUGHAN, W.E. (1969). *Rev. scient. Instrum.* **40**, 908.
CZERSKI, P. (1974). *Biologic effects and health hazards of microwave radiation.* Polish Medical Publishers, Warsaw.
DANFORD, M.D. and LEVY, H.A. (1962). *J. Am. chem. Soc.* **84**, 3965.
DAS, T.K. (1972). Ph.D. Thesis. University of Bradford.
DAS, T.K. and SMITH, R.B. (1972). *Proc. Int. Microwave Power Symposium.* Ottawa.
DANNHAUSER, W. (1970). *Rev. scient. Instrum.* **41**, 1110.
DAVIDSON, D.W., AUTY, R.P., and COLE, R.H. (1951). *Rev. scient. Instrum.* **22**, 678.
DAVIES, G.J. and CHAMBERLAIN, J. (1972). *J. Phys. A.* **5**, 767.
DAVIES, M. (1969). Chapter 5 of Hill *et al.* (1969).
DAVIES, M. (1965). *Some electrical and optical aspects of molecular behaviour.* Pergamon, Oxford.
DAVIS, F. and LOEB, H.W. (1965). *Proc. Instn elect. electron. Engrs* **53**, 1649.
DAWKINS, A.W.J., SHEPPARD, R.J., and GRANT, E.H. (1978). To be published.
DEBYE, P. (1929). *Polar molecules.* Reinhold, New York.
DEBYE, P. and FALKENHAGEN, H. (1928). *Phys. Z.* **29**, 121 and 401.
DEBYE, P. and HÜCKEL, E. (1923). *Phys. Z.* **24**, 185.

DEL BENE, J. and POPLE, J.A. (1970). *J. chem. Phys.* **52**, 4853.
DELBOS, G. and SALEFRAN, J.-L. (1972). *C.r. hebd. Séanc. Acad. Sci., Paris* **274**, 259.
DE LOOR, G.P. (1973). *J. Microwave Power* **8**, 67.
DE LOOR, G.P., VAN GEMERT, M.J.C., and GRAVESTEYN, H. (1973). *Chem. Phys. lett.* **18**, 295.
DRAPER, N.R. and SMITH, H. (1966). *Applied regression analysis*. Wiley, New York.
EDSALL, J.T. and WYMAN, J. (1958). *Biophysical chemistry*. Academic Press, New York.
EINOLF, C.W. and CARSTENSEN, E.L. (1969). *Biophys. J.* **9**, 634.
EINOLF, C.W. and CARSTENSEN, E.L. (1973). *Biophys. J.* **13**, 8.
EISENBERG, D. and KAUZMANN, W. (1969). *The structure and properties of water*. Clarendon Press, Oxford.
ESSEX, C.G. (1976). Ph.D. Thesis, University of London.
ESSEX, C.G., GRANT, E.H., SHEPPARD, R.J., SOUTH, G.P., SYMONDS, M.S., MILLS, G.L., and SLACK, J. (1977b). *Ann. N.Y. Acad. Sci.* (In press.)
ESSEX, C.G., SOUTH, G.P., SHEPPARD, R.J., and GRANT, E.H. (1975). *J. Phys. E.* **8**, 385.
ESSEX, C.G., SYMONDS, M.S., SHEPPARD, R.J., GRANT, E.H., LAMOTE, R., SOETEWEY, F., ROSSENEU, M.Y., and PEETERS, H. (1977a). *Phys. Med. Biol.* **22**, 1160.
EYRING, H., GLASSTONE, S., and LAIDLER, K.J. (1941). *Theory of rate processes*. McGraw-Hill, New York.
FALKENHAGEN, H. (1934). *Electrolytes*. Clarendon Press, Oxford.
FATUZZO, E. and MASON, P.R. (1967). *Proc. phys. Soc.* **90**, 741.
FELLNER-FELDEGG, H. (1969). *J. phys. Chem., Ithaca* **73**, 616.
FELLNER-FELDEGG, H. (1972). *J. phys. Chem., Ithaca* **76**, 2116.
FELLNER-FELDEGG, H. and BARNETT, E.F. (1970). *J. phys. Chem., Ithaca* **74**, 1962.
FERRIS, C.D. (1963). *Rev. scient. Instrum.* **34**, 109.
FERRY, J.D. and ONCLEY, J.L. (1941). *J. Amer. chem. Soc.* **63**, 272.
FINSEY, R. and VAN LOON, R. (1972). *High frequency dielectric measurements* (ed. J.E. Chamberlain and G.W. Chantry). IPC, London.
FISHER, H.F. (1965). *Biochim. biophys. Acta* **109**, 544.
FORSTER, E.O. and MINTON, A.P. (1972). *Phys. Methods macromol. Chem.* **2**, 185.
FRANK, H.S. (1972). In: *Water — a comprehensive treatise* (ed. F. Franks), Vol. 1, p. 515. Plenum, London.
FRANK, H.S. and WEN, W.Y. (1957). *Discuss. Faraday Soc.* **24**, 133.
FRANKS, F. (1972). *Water — a comprehensive treatise*, Vol. 1. Plenum, London.
FREDERICQ, E. and HOUSSIER, C. (1973). *Electric dichroism and electric birefringence*. Clarendon Press, Oxford.
FREEMAN, J. (1967). *Trans. N.Y. Acad. Sci.* **29**, 623.
FRICKE, H. (1925). *J. gen. Physiol.* **9**, 137.
FRICKE, H. (1955). *J. phys. Chem., Ithaca* **59**, 168.
FRICKE, H. and CURTIS, H.J. (1934). *Nature, Lond.* **133**, 651.
FRICKE, H. and CURTIS, H.J. (1937). *J. phys. Chem., Ithaca* **41**, 729.
FRICKE, H. and MORSE, S. (1926). *J. gen. Physiol.* **9**, 153.
FRICKE, H., SCHWAN, H.P., LI, K., and BRYSON, U. (1956). *Nature, Lond.* **177**, 134.
FRÖHLICH, H. (1958). *Theory of dielectrics*. Clarendon Press, Oxford.
GARG, S.K., KILP, H., and SMYTH, C.P. (1965). *J. chem. Phys.* **43**, 2341.
GARG, S.K. and SMYTH, C.P. (1965). *J. chem. Phys.* **43**, 2959.
GARNHAM, R.H. (1958). *R.R.E. Report No. 3020*, March 1958.
GENT, W.L.G., GRANT, E.H., and TUCKER, S.W. (1970). *Biopolymers* **9**, 124.
GIESE, K. and TIEMANN, R. (1975). *Adv. mol. Relax. Processes* **7**, 45.
GLARUM, S. (1960). *J. chem. Phys.* **33**, 1371.

GOEBEL, V.W. and VOGEL, H. (1964). *Z. Naturf.* **196**, 292.
GORDON, R.G., MALLARD, J.R. and PHILIP, J.F. (1976). In: *NMR and cancer* (ed. R. Damadian). Pacific Book Pub., Palo Alto, California.
GOSWAMI, D.N., DAS, J., and DAS GUPTA, D.D. (1973). *Biopolymers* **12**, 1047.
GRANT, E.H. (1956). Ph.D. Thesis, University of London.
GRANT, E.H. (1957a). *J. chem. Phys.* **26**, 1575.
GRANT, E.H. (1957b). *Phys. Med. Biol.* **2**, 17.
GRANT, E.H. (1959). *Br. J. appl. Phys.* **10**, 87.
GRANT, E.H. (1962). *Nature, Lond.* **196**, 1194.
GRANT, E.H. (1965). *Ann. N.Y. Acad. Sci.* **125**, 418.
GRANT, E.H. (1966). *J. mol. Biol.* **19**, 133.
GRANT, E.H. (1969). *J. phys. Chem., Ithaca* **73**, 4386.
GRANT, E.H., BUCHANAN, T.J., and COOK, H.F. (1957). *J. chem. Phys.* **26**, 156.
GRANT, E.H. and KEEFE, S.E. (1968). *Rev. scient. Instrum.* **39**, 1800.
GRANT, E.H., KEEFE, S.E., and SHACK, R. (1972). *Adv. mol. Relax. Processes* **4**, 217.
GRANT, E.H., KEEFE, S.E., and TAKASHIMA, S. (1968). *J. phys. Chem., Ithaca* **72**, 4373.
GRANT, E.H., MITTON, B.G.R., SOUTH, G.P., and SHEPPARD, R.J. (1974). *Biochem. J.* **139**, 375.
GRANT, E.H. and SHACK, R. (1967). *Br. J. appl. phys.* **18**, 1807.
GRANT, E.H. and SHACK, R. (1969). *Trans. Faraday Soc.* **65**, 1519.
GRANT, E.H. and SHEPPARD, R.J. (1970). *J. Phys. D.* **3**, 84.
GRANT, E.H. and SHEPPARD, R.J. (1974a). *J. chem. Phys.* **60**, 1792.
GRANT, E.H. and SHEPPARD, R.J. (1974b). *Phys. Med. Biol.* **19**, 153.
GRANT, E.H., SHEPPARD, R.J., MILLS, G.L., and SLACK, J. (1972). *Lancet* i, 1159.
GRANT, E.H., SHEPPARD, R.J., and SOUTH, G.P. (1975). *Proc. 5th European Microwave Conference*
GRANT, E.H., SOUTH, G.P., TAKASHIMA, S., and ICHIMURA, H. (1971). *Biochem. J.* **122**, 691.
GRANT, E.H., SOUTH, G.P., and WALKER, I.O. (1971). *Biochem. J.* **122**, 765.
HAGGIS, G.H., HASTED, J.B., and BUCHANAN, T.J. (1952). *J. chem. Phys.* **20**, 1452.
HAGGIS, G.H., MITCHIE, D., MUIR, A.R., ROBERTS, K.B., WALKER, P.M.B., BLOW, B.M., and SHEININ, R. (1974). *Introduction to Molecular Biology* (2nd ed.). Longman, London.
HALLENGA, K. (1972). Ph.D. Thesis, University of Gronigen.
HAMMING, R.W. (1962). *Numerical methods for scientists and engineers*. McGraw Hill, New York.
HANSS, M. (1966). *Biopolymers* **4**, 1035.
HANSS, M. and BANNERJEE, R. (1967). *Biopolymers* **5**, 879.
HANSS, M. and BERNENGO, J.C. (1973). *Biopolymers* **12**, 215.
HAR-KEDAR, I. and BLEEHEN, N.M. (1976). *Adv. Rad. Biol.* **6**, 229.
HARVEY, S.C. and HOEKSTRA, P. (1972). *J. phys. Chem., Ithaca* **76**. 2987.
HASTED, J.B. (1973). *Aqueous dielectrics*. Chapman and Hall, London.
HENDRICKX, H., VERBRUGGEN, R., ROSSENEU-MOTREFF, M.Y., BLATON, V., and PEETERS, H. (1968). *Biochem. J.* **110**, 419.
HILL, N.E. (1963). *Trans. Faraday Soc.* **59**, 344.
HILL, N.E. (1969). Chapter 1 of Hill *et al.* (1969).
HILL, N.E. (1972). *J. Phys. C.* **5**, 415.
HILL, N.E., VAUGHAN, W.E., PRICE, A.H., and DAVIES, M. (1969). *Dielectric properties and molecular behaviour*. Van Nostrand, London.
HÖBER, R. (1913). *Arch. ges. Physical.* **150**, 15.
HODGKIN, A.L. and KEYNES, R.D. (1953). *J. Physiol.* **119**, 513.
HORIKX, C.M. (1970). *J. Phys. E.* **3**, 871.
HYDE, P.J. (1970). *Proc. Instn elect. Engrs* **117**, 1891.
ISHIBASHI, Y., SAWADA, A., and TAKAGI, Y. (1969). *J. phys. Soc. Japan.* **27**, 705.

ISKANDER, M.F. and STUCHLY, S.S. (1972). *IEEE Trans.* **IM-21**, 425.
JACKSON, W. (1964). *High frequency transmission lines.* Methuen, London.
JACOBSON, B. (1953). *Nature, Lond.* **172**, 666.
JACOBSON, B. (1955). *J. Am. Chem. Soc.* **77**, 2919.
JASON, A.C. and LEES, A. (1971). *Dept of Trade and Industry, Torry Research Stn Report* T 71/7. H.M.S.O., London.
JOHNSON, C.C. and GUY, A.W. (1972). *Proc. IEEE.* **6**, 692.
JORDAN, B.P. (1971). Ph.D. Thesis, University of London.
JORDAN, B.P. and GRANT, E.H. (1970a). *J. phys. E.* **3**, 764.
JORDAN, B.P. and GRANT, E.H. (1970b). *J. phys. D.* **3**, 1068.
JUNGER, G., JUNGER, I., and ALLGEN, L.G. (1949). *Nature, Lond.* **163**, 849.
KEEFE, S.E. and GRANT, E.H. (1971). *Protides Biol. Fluids* **19**, 355. Pergamon, Oxford.
KEEFE, S.E., and GRANT, E.H. (1974). *Phys. Med. Biol.* **19**, 701.
KENDREW, J.C., DICKERSON, R.E., STRANDBERG, B.E., HART, R.G., and DAVIES, D.R. (1960). *Nature, Lond.* **185**, 422.
KENT, M. (1970). *J. Phys. D.* **3**, 1275.
KENT, M. (1972). *J. Phys. D.* **5**, 394.
KENT, M. (1977). *J. Microwave Power* **12**, 101.
KIRKWOOD, J.G. (1932). *J. chem. Phys.* **2**, 35.
KIRKWOOD, J.G. (1939). *J. chem. Phys.* **7**, 911.
KIRKWOOD, J.G. and SHUMAKER, J.B. (1952). *Proc. natn Acad. Sci. U.S.A.* **38**, 855.
KITTEL, C. (1971). *Elementary statistical physics*, p. 129. Wiley, New York.
KUBO, K. (1957). *J. phys. Soc. Japan* **12**, 570.
KUNTZ, I.D. and KAUZMANN, W. (1974). *Adv. Protein Chem.* **28**, 239.
LAMONT, H.R.L. (1959). *Waveguides.* Methuen, London.
LAMOTE, R., DENOO, A., ROSSENEU-MOTREFF, M.Y., and PEETERS, H. (1971). *Protides Biol. Fluids* **19**, 371. Pergamon, Oxford.
LAWINSKI, C.P. (1974). Ph.D. Thesis, University of London.
LAWINSKI, C.P., SHEPHERD, J.C.W., and GRANT, E.H. (1975). *J. Microwave Power* **10**, 147.
LIM, J.J. and SHAMOS, M.H. (1971). *Biophys. J.* **11**, 648.
LINDLEY, D.W. and MILLER, J.C.P. (1964). *Cambridge elementary statistical tables.* Cambridge University Press.
LING, G.N. (1972). In: *Water and aqueous solutions* (ed. R.A. Horne). Wiley, London.
LITTLE, V.I. and SMITH, V. (1955). *Proc. phys. Soc. B.* **68**, 65.
LOEB, H.W. (1972). *IEEE Trans.* **IM-21**, 166.
LOEB, H.W., YOUNG, G.M., QUICKENDEN, P.A., and SUGGETT, A. (1971). *Ber. Bunsenges. phys. Chem.* **75**, 1155.
LOVELL, S.E. and COLE, R.H. (1959). *Rev. scient. Instrum.* **30**, 361.
LUMRY, R. and YUE, R. (1965). *J. phys. Chem., Ithaca* **69**, 1162.
McTAGUE and GIBBS, J. (1966). *J. chem. Phys.* **44**, 4295.
MALMBERG, C.G. and MARYOTT, A.A. (1956). *J. Res. natn Bur. Stand.* **56**, 1.
MANDEL, M. (1961). *Mol. Phys.* **4**, 489.
MANDEL, M. (1965). *Protides Biol. Fluids* **13**, 415. Elsevier, Amsterdam.
MANDEL, M. (1977). *Ann. N.Y. acad. Sci.*(In press.)
MARCUVITZ, N. (1951). *Waveguide Handbook.* McGraw Hill, New York.
MARINO, A.A. (1967). *Phys. Med. Biol.* **12**, 367.
MARQUARDT, D.W. (1963). *J. Soc. ind. appl. Math.* **2**, 431.
MARQUARDT, D.W., BENNETT, R.G., and BURRELL, E.J. (1961). *J. molec. Spectrosc.* **7**, 269.
MAXWELL, J.C. (1892). *A treatise on electricity and magnetism.* Clarendon Press, Oxford.
MEDINA, D., HAZLEWOOD, C.F., CLEVELAND, C.G., CHANG, D.C., SPJUT, H.J., and MOYERS, R. (1975). *J. natn Cancer Inst.* **54**, 813.
MEREDITH, R. and PREECE, G.H. (1963). *IEEE. Trans.* **MTT-11**, 332.

MINAKATA, A. (1966). *Biochim. biophys. Acta* **126**, 570.
MINAKATA, A., IMAI, N., and OOSAWA, F. (1972). *Biopolymers* **11**, 347.
MINTON, A.P. (1971). *Nature, Lond.* **234**, 165.
MORGAN, S.P. (1949). *J. appl. Phys.* **20**, 352.
MOSER, P., SQUIRE, P.G., and O'KONSKI, C.T. (1966). *J. phys. Chem., Ithaca* **70**, 744.
MUNGALL, A.G. and HART, J. (1957). *Can. J. Phys.* **35**, 995.
NEALE, S.M. and WEYL, D.A. (1966). *Proc. R. Soc.* **A291**, 368.
NELDER, J.A. and MEAD, R. (1965). *Comput. J.* **7**, 308.
NELSON, S.O. (1977). *J. Microwave Power* **12**, 67.
NELSON, S.O. and CHARITY, L.F. (1972). *Trans. Am. Soc. agric. Engrs* **15**, 1099.
NEMETHY, G. and SCHERAGA, H.A. (1964). *J. chem. Phys.* **41**, 680.
NICOLSON, A.M. (1968). *IEEE Trans.* **IM-17**, 395.
NICOLSON, A.M. and ROSS, G.F. (1970). *IEEE Trans.* **IM-19**, 377.
NORDQVIST, P., ARWIN, H., and LUNDSTRÖM, I. (1975). *10th Int. Congress of Gerontology*, Jerusalem.
O'BRIEN, B.B. (1967). *IEEE Trans.* **IM-16**, 124.
O'KONSKI, C.T. (1960). *J. phys. Chem., Ithaca* **64**, 605.
O'KONSKI, C.T. and EDWARDS, A. (1968). *Rev. scient. Instrum.* **39**, 1456.
OLIVER, B.M. (1964). *Hewlett-Packard J.* **15**, 1.
ONCLEY, J.L. (1938). *J. Amer. chem. Soc.* **60**, 1115.
ONCLEY, J.L. (1942). *Chem. Rev.* **30**, 433.
ONCLEY, J.L. (1943). Chapter 22 of Cohn and Edsall (1943).
ONSAGER, L. (1931). *Phys. Rev.* **37**, 405.
ONSAGER, L. (1936). *J. Am. chem. Soc.* **58**, 1486.
OOSAWA, F. (1970). *Biopolymers* **9**, 677.
PAULING, L. (1960). *The nature of the chemical bond* (3rd ed.). Cornell, Ithaca, New York.
PAULING, L. (1961). *Science, N.Y.* **134**, 15.
PAULY, H. and PACKER, L. (1960). *J. biophys. and biochem. Cytol.* **7**, 603.
PAULY, H., PACKER, L., and SCHWAN, H.P. (1960). *J. biophys. and biochem. Cytol.* **7**, 589.
PAULY, H. and SCHWAN, H.P. (1964). *I.E.E.E. Trans. Bio-Med. Eng.* **BME-11**, 103.
PAULY, H. and SCHWAN, H.P. (1966). *Biophys. J.* **6**, 621.
PAYNE, R. and THEODOROU, I.E. (1971). *Rev. scient. Instrum.* **42**, 218.
PAYNE, R. and THEODOROU, I.E. (1972). *J. phys. Chem., Ithaca* **76**, 2892.
PENNOCK, B.E. and SCHWAN, H.P. (1969). *J. phys. Chem., Ithaca* **73**, 2600.
PERRIN, F. (1934). *J. Phys. Radium, Paris* **5**, 497.
PIMENTEL, G.C. and McCLELLAN, A.L. (1960). *The hydrogen bond*. Freeman, San Francisco.
PINE, C., ZOELLNER, W.G., and ROHRBAUGH, J.H. (1959). *J. opt. Soc. Am.* **49**, 1202.
PLIMPTON, S.J. and LAWTON, W.E. (1936). *Phys. Rev.* **50**, 1066.
POPLE, J.A. (1951). *Proc. R. Soc.* **A205**, 163.
PRICE, A.H. (1969). Chapters 3 and 4 of Hill *et al.* (1969).
RAHMAN, A. and STILLINGER, F.H. (1971). *J. chem. Phys.* **53**, 3336.
RAJEWSKY, B. and SCHWAN, H.P. (1948). *Naturwissenschaften* **35**, 315.
ROBERTS, A. (1977). M.Phil. Thesis, University of London.
ROBERTS, A., DAWKINS, A.W.J., SHEPPARD, R.J., and GRANT, E.H. (1978). To be published.
ROBERTS, S. and VON HIPPEL, A. (1946). *J. appl. Phys.* **17**, 610.
ROSEN, D. (1963). *Trans. Faraday Soc.* **59**, 2178.
ROSEN, D., BIGNALL, R., WISSE, J.D.M., and VAN DER DRIFT, A.C.M. (1969). *J. Phys. E.* **2**, 22.
ROSENBROCK, H.H. (1960). *Comput. J.* **3**, 175.
ROSSENEU-MOTREFF, M.Y., SOETEWEY, F., LAMOTE, R., and PEETERS, H. (1973). *Biopolymers* **12**, 1259.

RZEPECKA, M.A. and STUCHLY, S.S. (1975). *IEEE Trans.* **IM-24**, 27.
SALEFRAN, J.L., DELBOS, G., MARZAT, U., and BOTTREAU, A.M. (1977). *Advanc. mol. relax. proc.* **10**, 35.
SAMULON, H.A. (1951). *Proc. Instn radio Engrs* **39**, 175.
SANDUS, O. and LUBITZ, B.B. (1961). *J. phys. Chem., Ithaca* **65**, 881.
SAUNDERS, J.F. and FLYNN, J.E. (1964). Forms of water in biological systems, *Ann. N.Y. acad. Sci.* (1965), **125**, 249–772.
SCAIFE, B.K.P. (1971). *Complex permittivity*. English Universities Press, London.
SCHEIDER, W. (1965). *Biophys. J.* **5**, 617.
SCHERAGA, H.P. (1961). *Protein structure*. Academic Press, New York.
SCHLECHT, P. (1969). *Biopolymers* **8**, 757.
SCHLECHT, P., MAYER, A., HETTNER, G., and VOGEL, H. (1969). *Biopolymers* **7**, 963.
SCHLECHT, P., VOGEL, H., and MAYER, A. (1968). *Biopolymers* **6**, 1717.
SCHOENBORN, B.P., FEATHERSTONE, R.M., VOGELHUT, P.O., and SUSSKIND, C. (1964). *Nature, Lond.* **202**, 696.
SCHWAN, H.P. (1957). *Adv. biol. med. Phys.* **5**, 147.
SCHWAN, H.P. (1963). *Physical techniques in biological research*. Academic Press, New York.
SCHWAN, H.P. (1965). *Ann. N.Y. acad. Sci.* **125**, 344.
SCHWAN, H.P. (1974). In: *Biologic effects and health hazards of microwave radiation* (ed. P. Czerski), p. 152. Polish Medical Publishers, Warsaw.
SCHWAN, H.P. (1977). *Ann. N.Y. acad. Sci.* (In press.)
SCHWAN, H.P. and COLE, K.S. (1960). *Medical Physics* **3**, 152.
SCHWAN, H.P. and FOSTER, K.R. (1977). *Biophys. J.* **17**, 193.
SCHWAN, H.P. and LI, K. (1953). *Proc. Instn radio Engrs* **41**, 1735.
SCHWAN, H.P. and MACZUK, J. (1960). *Rev. scient. Instrum.* **31**, 59.
SCHWAN, H.P., SHEPPARD, R.J., and GRANT, E.H. (1976). *J. chem. Phys.* **64**, 2257.
SCHWAN, H.P. and SITTEL, K. (1953). *Trans. Am. Inst. elect. Engrs* **5**, 114.
SCHWAN, H.P., TAKASHIMA, S., MIYAMOTO, V.K., and STOEKENIUS, W. (1970). *Biophys. J.* **10**, 1102.
SCHWARZ, G. (1962). *J. phys. Chem., Ithaca* **66**, 2636.
SCHWARZ, G. (1972). *Adv. Mol. Relax. Processes* **3**, 281.
SHACK, R. (1972). Ph.D. Thesis, University of London.
SHANNON, C.E. (1949). *Proc. Instn radio Engrs* **37**, 10.
SHEPHERD, J.C.W. (1967). Ph.D. Thesis, University of London.
SHEPHERD, J.C.W. and GRANT, E.H. (1968a). *Proc. R. Soc.* **A307**, 345.
SHEPHERD, J.C.W. and GRANT, E.H. (1968b). *Proc. R. Soc.* **A307**, 335.
SHEPPARD, R.J. (1971). Ph.D. Thesis, University of London.
SHEPPARD, R.J. (1972). *J. Phys. D.* **5**, 1576.
SHEPPARD, R.J. and GRANT, E.H. (1972). *J. Phys. E.* **5**, 1208.
SHEPPARD, R.J. and GRANT, E.H. (1974). *Adv. Mol. Relax. Processes* **6**, 61.
SHEPPARD, R.J., JORDAN, B.P., and GRANT, E.H. (1970). *J. Phys. D.* **3**, 1759.
SIMHA, R. (1940). *J. phys. Chem., Ithaca* **44**, 25.
SOUTH, G.P. (1970). Ph.D. Thesis, University of London.
SOUTH, G.P. and GRANT, E.H. (1972). *Proc. R. Soc.* **A328**, 371.
SOUTH, G.P. and GRANT, E.H. (1973). *Biopolymers* **12**, 1937.
SOUTH, G.P. and GRANT, E.H. (1974). *Biopolymers* **13**, 1777.
SPENDLEY, W., HEXT, C.R., and HIMSWORTH, F.C. (1962). *Technometrics* **4**, 441.
STARR, A.T. (1932). *Wireless Engr exp. Wireless* **9**, 615.
STRATTON, J.A. (1941). *Electromagnetic theory*. McGraw Hill, New York.
STUMPER, U. (1973). *Rev. scient. Instrum.* **44**, 165.
SUGGETT, A. (1975). *J. Phys. E.* **8**, 327.
SZWARNOWSKI, S. and SHEPPARD, R.J. (1977). *J. Phys. E.* **10**, 1163.
TAIT, M.J., SUGGETT, A., FRANKS, F., ABLETT, S., and QUICKENDEN, P.A. (1972). *J. solution Chem.* **1**, 131.

TAKASHIMA, S. (1962). *J. Polym. Sci.* **56**, 257.
TAKASHIMA, S. (1963). *J. Polym. Sci. Part A* **1**, 2791.
TAKASHIMA, S. (1965). *Biopolymers* **4**, 663.
TAKASHIMA, S. (1966). *J. phys. Chem., Ithaca* **70**, 1372.
TAKASHIMA, S. (1967a). *Biopolymers* **5**, 899.
TAKASHIMA, S. (1967b). *Adv. Chem.* **63**, 232.
TAKASHIMA, S. (1973). *Biopolymers* **12**, 145:
TAKASHIMA, S. and LUMRY, R. (1958a). *J. Am. chem. Soc.* **80**, 4238.
TAKASHIMA, S. and LUMRY, R. (1958b). *J. Am. chem. Soc.* **80**, 4244.
TAKASHIMA, S. and SCHWAN, H.P. (1965). *J. phys. Chem., Ithaca* **69**, 4176.
TAUB, J.J., HINDIN, M.J., HICKELMANN, O.F., and WRIGHT, M.L. (1963). *IEEE Trans.* **MTT-11**, 338.
TOMASELLI, V.P. and SHAMOS, M.H. (1973). *Biopolymers* **12**, 353.
VAN DER TOUW, F., BRIEDÉ, J.W.H., and MANDEL, M. (1973). *Biopolymers* **12**, 111.
VAN DER TOUW, F. and MANDEL, M. (1971). *Trans. Faraday Soc.* **67**, 1336.
VAN DER TOUW, F., MANDEL, M., HONIJK, D.D., and VERHOOG, H.G.F. (1971). *Trans. Faraday Soc.* **67**, 1343.
VAN GEMERT, M.J.C. (1971). *J. phys. Chem. Ithaca* **75**, 1323.
VAN GEMERT, M.J.C. (1973). *Philips Res. Rep.* **28**, 530.
VAN GEMERT, M.J.C. (1974a). *J. chem. Phys.* **60**, 3963.
VAN GEMERT, M.J.C. (1974b). *Adv. Mol. Relax. Processes* **6**, 123.
VAN GEMERT, M.J.C. and BORDEWIJK, P. (1972). *Appl. scient. Res.* **27**, 156.
VAN GEMERT, M.J.C. and DE GRAAN, J.G. (1972). *Appl. scient. Res.* **26**, 1.
VAN LOON, R. and FINSEY, R. (1973). *Rev. scient. Instrum.* **44**, 1204.
VAN LOON, R. and FINSEY, R. (1974). *Rev. scient. Instrum.* **45**, 523.
VAN LOON, R. and FINSEY, R. (1975). *J. Phys. D.* **8**, 1232.
VAUGHAN, W.E., BERGMANN, K., and SMYTH, C.P. (1961). *J. phys. Chem., Ithaca* **65**, 94.
VON CASIMIR, W., KAISER, N., KEILMANN, F., MAYER, A., and VOGEL, H. (1968). *Biopolymers* **6**, 1705.
VOSS, W.A.G., RAJOTTE, R.V., and DOSSETOR, J.B. (1974). *J. Microwave Power* **9**, 181.
WADA, A. (1959). *J. chem. Phys.* **31**, 495.
WADA, A. (1960). *Bull. chem. Soc. Japan.* **33**, 822.
WAGNER, K. (1913). *Annln Physik.* **40**, 817.
WAGNER, K. (1914). *Arch. Elektrotech.* **2**, 371.
WALRAFEN, G.E. (1968). In: *Equilibria and reaction kinetics in hydrogen bonded solvent systems* (ed. A.K. Covington). Taylor and Francis, London.
WANG, J.H. (1965). *J. phys. Chem., Ithaca* **69**, 4412.
WATSON, J.D. and CRICK, F.H.C. (1953). *Nature, Lond.* **171**, 734.
WEXLER, A. (1969). *IEEE Trans.* **MTT-17**, 416.
WHITTINGHAM, T.A. (1970). *J. phys. Chem., Ithaca* **74**, 1824.
WILLIAMS, G. (1959). *J. phys. Chem., Ithaca* **63**, 534.
WYMAN, J. (1936). *Chem. Rev.* **19**, 213.
WYMAN, J. and INGALS, E.N. (1943). *J. biol. Chem.* **147**, 297.
YOUNG, S.E. (1967). Ph.D. Thesis, University of London.
YOUNG, S.E. and GRANT, E.H. (1968). *J. Phys. E.* **1**, 429.
ZAFAR, M.S., HASTED, J.B., and CHAMBERLAIN, J. (1973). *Nature (Phys. Sci.), Lond.* **243**, 106.

AUTHOR INDEX

Aaron, M.W., 176, 177, 183
Ablett, S., 180
Adey, W.R., 22
Allgen, L.G., 215
Allison, J., 93
Arwin, H., 19
Asami, K., 219
Asfar, M.N., 145, 155
Auty, R.P., 101, 120

Baden Fuller, A.J., 80, 94, 95
Bannerjee, R., 206
Barbenza, G.H., 85
Barnett, E.F., 102, 113
Bassett, H.L., 100
Bateman, J.B., 7
Bather, R., 21
Bawin, S.M., 22
Bennett, R.G., 125, 127
Benson, F.A., 93
Berberian, J.G., 66
Berendsen, H.J.C., 14
Bergmann, K., 100, 143
Bernal, J.D., 151
Bernengo, J.C., 214
Bignall, R., 62, 69
Bjerrum, N., 159
Blaton, V., 194, 201, 203
Bleaney, B., 45
Bleaney, B.I., 45
Bleehen, N.M., 20
Blow, B.M., 9, 11
Bordewijk, P., 102, 113
Bottreau, A.M., 187
Brehm, G.A., 113
Brey, W.S., 171
Briedé, J.W.H., 216
Broadhurst, M.G., 63
Brophy, J.J., 101
Brownlee, K.A., 131
Bryson, U., 219
Bucci, O.M., 113
Buchanan, T.J., vi, 7, 82, 85, 86, 94, 123, 151, 152, 153, 155, 157, 167, 187, 189
Buckingham, A.D., 184
Bur, A.J., 63
Burge, R.E., vi
Burger, M.M., 21
Burrell, E.J., 125, 127

Calvert, R., 61

Carson, J.W. 100
Carstensen, E.L., 19, 219
Catchpool, J.F., 18
Chamberlain, J., 58, 100, 155
Chang, D.C., 21
Charity, L.F., 21
Chisholm, J.S.R., 52
Clark, A.H., 102, 114, 115, 117, 118
Clarke, R.P., 182
Clarkson, T.S., 119
Clegg, J.S., 20
Cleveland, C.G., 21
Cohn, E.J., 174, 175
Cole, K.S., 122, 217
Cole, R.H., 52, 61, 66, 101, 118, 120, 122, 218
Collie, C.H., 100
Conner, W.P., 7, 177, 182, 184
Cook, H.F., 7, 20, 123, 219
Cooley, J.W., 113
Cortucci, G., 113
Coulson, C.A., 151
Cox, M.A., 219
Crick, F.H.C., 13, 14
Croom, E.J., 178, 185
Cullen, A.L., 100
Cunningham, T.A., 21
Curtis, H.J., 65, 219
Cutnell, J.D., 98
Czerski, P., 22

Danford, M.D., 151
Dannhauser, W., 62
Das, J., 215
Das, T.K., 20
Das Gupta, D.D., 215
Davidson, D.W., 101, 120
Davies, D.R., 198
Davies, G.J., 100
Davies, M., 26, 61, 146
Davis, F., 101
Dawkins, A.W.J., 11, 119
Debye, P., 1, 34, 157, 191, 193
de Graan, J.G., 102, 113
Del Bene, J., 154
Delbos, G., 187
de Loor, G.P., 21, 102
Denoo, A., 62, 74
Dickerson, R.E., 198
Dossetor, J.B., 21
Draper, N.R., 125, 127, 131, 135
Dutuit, Y., 187

Edsall, J.T., 174, 175, 182
Edwards, A., 62
Einolf, C.W., 219
Eisenberg, D., 144, 151, 154, 160
Essex, C.G., 17, 19, 71, 72, 73, 75, 162, 164, 215, 216
Eyring, H., 50

Falkenhagen, H., 91, 193
Fatuzzo, E., 52
Featherstone, R.M., 18
Fellner-Feldegg, H., 101, 102, 111, 112, 113, 114, 115, 118, 119
Ferris, C.D., 66
Ferry, J.D., 66
Finsey, R., 95, 97, 98, 99, 144
Fisher, H.F., 169
Flynn, J.E., 13
Forster, E.O., 16
Foster, K.R., 170
Fowler, R.H., 151
Franceshetti, G., 113
Frank, H.S., 151, 153, 158, 180
Franks, F., 13, 144, 151, 155
Fredericq, E., 190, 196
Freeman, J., 171
Fricke, H., 65, 167, 217, 219
Fröhlich, H., 40, 41, 42, 55, 56

Gallagher, J.C., vii
Garg, S.K., 100, 159
Garnham, R.H., 100
Gent, W.L.G., 166
Gibbs, J.H., 214
Giese, K., 102, 114
Glarum, S., 52
Glasser, L., 119
Glasstone, S., 56
Goebel, V.W., 194
Gordon, R.G., 21
Goswami, D.N., 215
Grant, E.H., 7, 11, 21, 64, 65, 71, 72, 73, 75, 82–87, 92, 95, 119, 123, 128, 143, 145, 146, 153, 157–159, 162–167, 169, 171, 175–181, 183, 185, 189, 193–195, 197, 205, 206, 208, 209, 211, 212, 215, 216, 219
Grant, K.M., vii
Gravesteyn, H., 102
Gross, P.M., 61
Guy, A.W., 20

Haggis, G.H., 7, 8, 9, 151, 152, 153, 155, 157, 167, 187, 189
Haigh, J., 58
Hallenga, K., 100

Hamming, R.W., 106
Hanai, T., 219
Hanss, M., 206, 214
Har-Kedar, I., 20
Hart, J., 95
Hart, R.G., 198
Harvey, S.C., 171
Hasted, J.B., 7, 100, 144, 145, 146, 151, 152, 153, 155, 157, 158, 159, 167, 187, 189
Hazlewood, C.F., 21
Heeb, M.A., 171
Hendrickx, H., 194, 201, 203
Hettner, G., 195
Hext, C.R., 125
Hickelmann, O.F., 100
Hill, N.E., 26, 40, 47, 52, 53, 61, 146, 155
Himsworth, F.C., 125
Hindin, M.J., 100
Hine, M.J., 58
Höber, R., 219
Hodgkin, A.L., 219
Hoekstra, P., 171
Honijk, D.D., 66
Horikx, C.M., 99
Houssier, C., 190, 196
Hückel, E., 193
Hyde, P.J., 120

Ichimura, H., 167, 195, 206
Imai, N., 214
Ingals, E.N., 205
Ishibashi, Y., 101
Iskander, M.F., 119

Jackson, W., 91
Jacobson, B., 14, 192
Jason, A.C., 19
Johnson, C.C., 20
Jordan, B.P., 71, 87, 128, 142, 181
Junger, G., 215
Junger, I., 215

Kaiser, N., 97
Kauzmann, W., 16, 17, 144, 151, 154, 160
Keefe, S.E., 82, 165, 167, 175, 180, 183
Keilmann, F., 97
Kendrew, J.C., 197
Kent, M., 19, 171
Keynes, R.D., 219
Kilp, H., 100
Kirkwood, J.G., 42, 150, 191, 192, 211
Kittel, C., 40
Koizumi, N., 219

AUTHOR INDEX

Kranbuehl, D.E., 98
Kubo, K., 51
Kuntz, I.D., 16, 17

Lacroix, Y., 187
Laidler, K.J., 50
Lamont, H.R.L., 76, 80, 94
Lamote, R., 17, 62, 74, 206
Lawinski, C.P., 84, 176, 177, 178, 179, 185
Lawton, W.E., 26
Lees, A., 19
Levy, H.A., 151
Li, K., 20, 78, 219
Lim, J.J., 171
Lindley, D.W., 131, 137, 138
Ling, G.N., 16
Little, V.I., 91
Loeb, H.W., 101, 102, 107, 108, 109, 110, 111, 112, 113
Lovell, S.E., 61
Lubitz, B.B., 176
Lumry, R., 74, 194, 206
Lundström, I., 19

McClellan, A.L., 150
McTague, J.P., 214
Maczuk, J., 62, 65
Mallard, J.R., 21
Malmberg, C.G., 93, 145, 153
Mandel, M., 62, 63, 65, 66, 214, 216
Marcuvitz, N., 80
Marino, A.A., 171
Marquardt, D.W., 125, 127, 128
Marquis, R.E., 219
Maryott, A.A., 93, 145, 153
Marzat, C., 187
Mason, P.R., 52
Maxwell, J.C. 45, 167, 192
Mayer, A., 97, 195, 206
Mead, R., 125
Medina, D., 21
Mercer, W.B., 219
Meredith, R., 93
Migchelsen, C., 14
Miller, J.C.P., 131, 137, 138
Mills, G.L., 162, 164, 166, 215, 216
Minakata, A., 194, 214
Minton, A.P., 16, 159
Miyamoto, V.K., 219
Mitchie, D., 9, 11
Mitton, B.G.R., 162, 167, 169, 171
Moreau, J.M., 187
Morgan, S.P., 93
Morris, R.M., 52
Morse, S., 219

Moser, P., 194, 202
Moyers, R., 21
Muir, A.R., 9, 11
Mungall, A.G., 95

Nagenthiram, P., 100
Natale, L.A., 219
Neale, S.M., 215
Nelder, J.A., 125
Nelson, S.O., 21
Nemethy, G., 152
Nicolson, A.M., 101, 119
Noonan, K.D., 21
Nordqvist, P., 19

O'Konski, C.T., 61, 192, 193, 194, 202, 215, 216
Oliver, B.M., 101
Oncley, J.L., 7, 8, 62, 66, 167, 190
Onsager, L., 52, 52, 191
Oosawa, F., 214, 216

Packer, L., 218, 219
Pauling, L., 18, 150, 153
Pauly, H., 19, 62, 216, 218, 219
Payne, R., 62, 78, 84
Peacocke, A.R., vii
Peeters, H., 17, 62, 74, 194, 201, 203, 206
Pennock, B.E., 86, 164, 167, 169
Perrin, F., 199
Philip, J.F., 21
Pimental, G.C., 150
Pine, C., 100
Plimpton, S.J., 26
Pople, J.A., 151, 154, 155
Potapenko, G., 7
Preece, G.H., 93
Price, A.H., 26, 61, 146

Quickenden, P.A., 102, 107, 108, 109, 110, 111, 112, 114, 115, 117, 118, 180

Rahman, A., 155, 157
Rajewsky, B., 219
Rajotte, R.U., 21
Ritson, D.M., 100
Roberti, D.M., 143
Roberts, A., 11
Roberts, K.B., 9, 11
Roberts, S., 78
Robinson, B.G., 7, 167, 189
Rohrbaugh, J.H., 100
Rosen, D., 62, 69, 169, 170, 171
Rosenbrock, H.H., 125
Ross, G.F., 101, 119
Rosseneu, M.Y., 17, 62, 74, 194, 201,

Rosseneu, M.Y. (*cont.*)
 203, 206
Rzepecka, M.A., 119

Salefran, J.L., 187
Samulon, H.A., 112
Sandus, O., 226
Saunders, J.F., 13
Savarese, C., 113
Sawada, A., 101
Scaife, B.K.P., 26
Scheider, W., 212
Scheraga, H.A., 152, 201
Schoenborn, B.P., 18
Schlecht, P., 195, 206, 209, 213
Schwan, H.P., vi, 3, 4, 7, 19, 20,
 22, 62, 63, 65, 78, 86, 146,
 162, 163, 164, 167, 169, 170,
 171, 216, 218, 219
Schwarz, G., 192, 193, 207, 214
Seaman, M.S., 93
Shack, R., 95, 96, 159, 176, 177,
 178, 180, 183, 187
Shamos, M.H., 171
Shannon, C.E., 112
Sheinin, R., 9, 11
Shepherd, J.C.W., 84, 85, 176, 177,
 178, 179, 181, 185
Sheppard, R.J., vi, 11, 17, 21, 71,
 72, 73, 74, 75, 82, 83, 86, 88,
 89, 90, 92, 99, 119, 128, 143,
 145, 146, 153, 158, 159, 162,
 164, 166, 167, 169, 171, 178,
 215, 216, 219
Shumaker, J.B., 192, 211
Simha, R., 201
Sittel, K., 7
Slack, J., 162, 164, 166, 215, 216
Smith, H., 125, 127, 131, 135
Smith, R.B., 20
Smith, V., 91
Smyth, C.P., 7, 100, 143, 159, 177,
 182, 184
Soetewey, F., 17, 206
South, G.P., vi, 21, 71, 72, 73,
 74, 75, 162, 164, 166, 167, 169,
 171, 175, 181, 193, 194, 195,
 197, 205, 206, 208, 209, 211,
 212, 215, 216
Spendley, W., 125
Spjut, H.J., 21
Squire, P.G., 194, 202
Starr, A.T., 61
Stillinger, F.H., 155, 157
Stoeckenius, W., 219
Stockmayer, W.H., 113
Strandberg, B.E., 198
Stratton, J.A., 76, 80, 92

Stuchly, S.S., 119
Stumper, U., 99
Suggett, A., 102, 107, 108, 109,
 110, 111, 112, 114, 115, 117,
 118, 180
Susskind, C., 18
Symonds, M.S., 17, 162, 164, 166,
 215, 216
Szwarnowski, S., 99, 144

Tait, M.J., 180
Takagi, Y., 101
Takashima, S., vi, 63, 66, 165, 167,
 171, 194, 195, 206, 214, 215,
 219
Taub, J.J., 100
Theodorou, I.E., 62, 78, 84
Tiberie, R., 113
Tiemann, R., 102, 114
Tomaselli, V.P., 171
Tucker, S.W., 166
Tukey, J.W., 113
Turner, E.M., 98
Tuxworth, R.W., 119

van der Drift, A.C.M., 62, 69
van der Touw, F., 63, 65, 66, 216
van Gemert, M.J.C., 102, 112, 113,
 114, 115, 116, 118, 119, 120
van Loon, R., 95, 97, 98, 99, 144
Vaughan, W.E., 61, 98, 100, 146
Verbruggen, R., 194, 201, 203
Verhoog, H.G.F., 66
Vogel, H., 97, 194, 195, 206
von Casimir, W., 97
von Hippel, A., 78
Vogelhut, P.O., 18
Voss, W.A.G., 21

Wada, A., 211
Wagner, K., 192
Walker, I.O., 209
Walker, P.M.B., 9, 11
Walfafen, G.E., 158
Wang, J.H., 156
Ward, T.M., 171
Watson, J.D., 13, 14
Webb, S.J., 21
Webb, S.L., 101
Wen, W.Y., 158
Wexler, A., 81
Weyl, D.A., 215
Whittingham, T.A., 102
Williams, A.D., 100
Williams, G., 119, 123
Wisse, J.D.M., 62, 69
Wright, M.L., 100
Wyman, J., 7, 8, 174, 182, 205

Young, G.M., 102, 107, 108, 109, 110, 111, 112
Young, S.E., 64, 65, 176
Yu, P.K., 100

Yue, R., 74, 194

Zafar, M.S., 155
Zoellner, W.G., 100

SUBJECT INDEX

activation
  enthalpy, 50, 163
  process, 50
adenosine triphosphate, see ATP
alanine, 178
  α-, 10, 176, 177, 182, 187
  β-, 10, 176, 177, 182, 184, 187
alanylglycine, 10, 115, 176, 177, 184
albumin
  bovine serum 162, 165, 180
    denaturation of delipidated and relipidated, 206
  egg, 162, 171, 194
amino acids, 9, 171–88
γ-aminobutyric acid, 187
ε-aminocaproic acid, 10, 176, 177, 182, 187
anaesthetic agents, 18
applications of dielectric studies, biological and medical, 17–22
aqueous solution of haemoglobin, dielectric disperson curve of, 3
Arrhenius plot for pure water, 147
ATP, 11
attenuation coefficient, 76, 91, 94, 97
axial ratio
  dependence of relaxation time upon, 200, 203

bacteria
  E. coli, 19, 219
  micrococcus, 219
birefringence, 196
blood, 219
  structure of lipoproteins in, 18, 166, 215
bone, 171
bound water, 5, 15, 16, 144, 180, 197
  in bovine serum albumin, 164, 165
  dielectric properties of, 160–7
  in egg albumin, 162
  in haemoglobin, 162, 163
  in lens, 22
  in myoglobin, 162, 169
  serum lipoprotein, and genetically determined heart disease, 19, 166
  in solutions of small biological molecules, 23
bovine serum albumin see under albumin
bridge
  microwave, 85
  measurements, high frequency, 70–5
  low frequency, electrode polarization in, 62–6
bridges, 7, 57
  dielectric, 58–75
Brownian
  motion, 37
  rotation, 53, 55

cancer
  treatment of by hyperthermia produced by radio waves and microwaves, 20, 21
cataract, microwave induced, 22
cavity resonator, 100
cell membranes, 4, 5, 19, 217–19
cells, dielectric measuring
  coaxial line, 82, 83
  for measurements in the time domain, 112
  suitable for measurements up to 50 MHz, 72
  suitable for measurements up to 100 MHz, 72
  waveguide, 95, 98
cells, living, 217–20
charge transfer mechanism, 184
chemical rate processes
  dielectric measurements in, 207
cholesterol, 14
coaxial
  cell, 61, 82–4
  line cell, 76
  lines, 57, 75–93, 80
  line techniques, 75, 77
Cole–Cole
  plot, 123
  function, 140, 141
collagen, 14, 171
complex propagation constant, 76
confidence contours, 134–6
confidence intervals, 130–3, 140
convolution integral, 141
correlation
  coefficients, 133–4, 140, 179
  function, 51–3
  microscopic, 53

SUBJECT INDEX

molecular, 53
parameter, 149, 155
time, 158, 163
counterions, 214
relaxation of, 189, 216

data analysis, computerized, 9, 121, 124–43
statistical, 130–47
Debye, 2
dispersion, 48
equations, 48–9
unit, 2, 33
decay function, 47, 53
dielectric
bridges, 58–75
decrement, 165–7
increment, 54
mixture equations, 168
relaxation, 35, 43, 44–7
studies, biological and medical applications, 17–22
dielectrics
polar, 32–4
non-polar, 32, 34–5
diglycine, 176
dipolar rotation, 34, 196
molecular, 34
theory of, 26
dipole moment, 2, 32, 34
calculated and experimental, whale and horse myoglobin, 211
calculation of, 209, 213
α-helix, 211
mean square, 43
permanent, 37
protein molecule, 8
dispersion
α-, 3, 4
β-, 3, 4, 16, 89
in protein solutions, 191–5
δ-, 5, 12, 162–5
γ-, 3, 5
dispersion curve, of system with widely differing relaxation times, 54
dispersion equations, 47–50
distortion polarization see polarization, induced
distribution of relaxation times, 55–6, 123
DNA, 6, 23, 213, 215
-proflavine complexes, 215
D_2O, 159

E. coli bacterium, 19, 219
egg albumin, 162, 171, 194
electric displacement vector, 38–9

erythrocyte, 219
ethane diol, 175, 181

familial hyperbetalipoproteinaemia, 216
F-distribution, 136–8
formamide, 181
frequency dependent permittivity, theory of, 50–8
frequency domain
measurements, 57
techniques, 99–101

Gaussian distribution, 141
globular protein, 22
glucose, 180
glycine, 10, 172, 176, 177, 178, 182, 183
dielectric dispersion curve of in water, 173
solution, variation of static permittivity of, 183
glycylalanine, 10, 176, 177, 184

haemoglobin, 2, 6, 18, 163, 164, 167, 170, 195, 206
α-helix, 11
dipole moment, 211
helix–coil transition, 23, 215
high-loss liquid, 90, 100
higher modes, 80–2
formation of, 81
hydration, 6, 170 (*see also* bound water)
critical, 171
hydrogen bond, 150
hyperthermia, 20
external beam, 21

ice, 144, 154, 159
insect control with electromagnetic waves, 21
ion atmosphere, 193
ionic conductivity
corrections for, 91–3, 95
isoelectric range of amino acid solutions, 184

Kirkwood correlation parameter, 42

least-squares
analysis, 124
curve fitting, 121, 141
minimization, 124–30
lens
material, 19, 219
water in, 22
lipid, 19

lipid (cont.)
  bilayer model, 19, 216
  core model, 19, 216
lipoproteins, 3, 18, 23, 215–16
  LDL, 162, 215
  molecule, 19
  structure of in blood serum, 18
lysozyme, 171

materials, polar and non-polar, 32
Maxwell's equations, 76, 80
Maxwell–Wagner
  effect, 4, 89
  mechanism, 192
membranes, 4, 19, 217–220
methanol, 181
micrococcus, 219
microwave
  bridge, 85, 89
  hazards, 21
  oven, 22
  therapy, 20
microwaves
  behavioural effects, 22
  hazards of, 21–2
  in treatment of malignant disease, 20
  to warm frozen organs, 21
mitochondria, 219
mixture theory, 161
mode
  dominant, 94
  filtering, 82
  TEM, 94
molecular
  dipole moment, 23
  rotation, 34–5
  shape in solution, 7, 22
multiple reflections, 103, 106
multiple response methods, 113–18
muscle tissue, 3, 170, 190
  relative permittivity of, 4
myelin, 166
myoglobin, 6, 115, 161, 167, 168, 169, 170, 172, 189, 193, 195, 196, 198
  relaxation time and concentration of, 197
  plot of axial ratio against molecular volume for, 204

non-ionizing radiation, hazards of, 22
nuclear magnetic resonance (NMR), 16
nucleic acids, 3
  DNA, 213–15

parallel plate cell, 62

peptides, 10, 171–88
permanent dipole
  moment, 37
  rotation, 191
permittivity
  cell for medium radio frequencies, 67
  complex, 43–57
  frequency-dependent, 50–8
  phase constant, 76, 77
phospholipid vesicles, 219
α-plot, 86, 91
polarizability, 34–5
polarization, 29, 32, 35, 44
  electrode, 62, 190
    four terminal bridge in, 66
    in low-frequency bridge measurements, 62–6
  induced, 41, 53
polar side chains, rotation of, 164
power absorption, difference in relaxation and in resonance, 44
principal mode (of propagation of electromagnetic waves), 78, 80
proline, 14, 176, 180, 181, 184, 186
propagation constant, 77, 82, 92, 95
proteins, 189–220
  globular, 22
  powders, 171
  solutions, β-dispersion in, 191–5
proton
  fluctuation, 164, 192, 195, 212
    contribution, 212
  times, 36
  jumping, 212

Q-factor, 99

radio waves in treatment of malignant disease, 20
relaxation time, 2, 10
  and axial ratio, 200
  and molecular weight, 185
  and viscosity
    for proteins, 195–6
    for pure water, 157
    for triglycine in water, 185
resonance, 43–4
ribonuclease, 170
rotary diffusion constants, 196

sample holder, 61
safety levels, in use of microwaves, 22
self inductance, 62, 70–5
  calibration procedure to correct, 75
specific increment, 207

SUBJECT INDEX

standing wave, 78–9, 85, 94
static permittivity, 36–8
  of fluids, theory of, 39–43
  of water, 149–56
static relative permittivity, 36–43
static susceptibility, 39
superposition theorem, 47, 53
surface conductivity, 192–3
systematic errors, 133

t-distribution, 130–2, 137,
TDS see Time Domain Spectroscopy
tendon, 171
thiourea, 175
time domain measurements see Time Domain Spectroscopy
Time Domain Reflectometry see Time Domain Spectroscopy
Time Domain Spectroscopy, 9, 58, 101–21, 180
  alternative methods, 107–18
  cell suitable for single response methods, 112
  for multiple response methods, 114
  reflection system, 109
  thin cell method, 118
  time referencing trace, 108
  transmission system, 110
time referencing
  TDS trace, 108
transmission
  line methods, 7
  lines, 57, 75
  measurements in time domain, 110
travelling wave, 79, 85–6, 89–91, 95
  automated method, 89
    circuit 89
  circuit, 90
triglycine, 175, 176, 177, 178, 179, 185, 186
  dielectric dispersion curve of in water, 179

urea, 7, 175, 180, 183, 188

valine, 178
  δ-, 187
variance–covariance matrix, 130, 133
virus, alfalfa mosaic, 216
viruses, 216
viscosity, 181
  intrinsic, 201–2, 205
  and relaxation time
    for proteins, 194–5
    for pure water, 157
    for triglycine in water, 185

wall loss
  corrections for, 91–3, 95
  calculated, 93
water, 2, 9, 13, 115, 144–89
  in *Artemia salina*, 20
  in biological systems, 12–17
    role of dielectric measurements in the study of, 12
  dielectric dispersion curve of, 145–56
  diffusion coefficient of, 156, 160
  in human body, 12
  of hydration see bound water
  in lens, 22
  molecule, 149
  viscosity of, 156, 160
waveguide bridge circuit, 95–6
waveguides, 57, 93–9

zwitterion, 9, 10, 180, 182

ציון